CHEMISTRY: A MODERN INTRODUCTION FOR SIXTH FORMS AND COLLEGES

PART 1

General and Inorganic Chemistry

Second edition

T. F. Chadwick, B.Sc., C.Chem., F.R.I.C., Dip. Chem. Eng.
College of Technology,
Reading

London
GEORGE ALLEN & UNWIN
Boston   Sydney

First published in 1972
Second edition 1977
Third impression 1981

GEORGE ALLEN & UNWIN LTD
40 Museum Street, London WC1A 1LU

© George Allen & Unwin (Publishers) Ltd 1971, 1977

ISBN 0 04 546003 5

Printed in Great Britain
in 10 on 11 Point Times Roman
by William Clowes (Beccles) Limited, Beccles and London

PREFACE TO THE SECOND EDITION

Such has been the progress in recent years in both the content and the comprehension of chemistry, that changes in the approach and presentation of the subject have become inevitable. Modern syllabuses call for a greater awareness of general principles and understanding, rather than a wide knowledge of factual detail. This book attempts to give a detailed and clear account of basic theoretical principles and modern concepts required for 'A' and 'S' level and for National Certificate in Sciences courses, without overloading the student with unnecessary detail.

The first part of the book is devoted to general principles. The results of the modern quantum theory and the importance of ionization energies in relation to electronic configuration are discussed in some detail. Covalent bonding is considered in terms of both hybridization and (at a later stage in the book) electron pair repulsion, while the importance of energetics in relation to ionic bonding is stressed. The structures of a range of crystalline types is included. Concepts such as electronegativity, oxidation state, free energy changes and redox potentials are explained and the integration of these concepts in subsequent chapters of the book dealing with the chemistry of the elements and their compounds is emphasized.

The chemistry of the elements is given under the broad headings of s-block, p-block and d-block elements in the second part of the book. Group trends are considered and a special introductory section on the complexes formed by the transition elements is included. Much of the factual material is presented in the form of charts and tables, and it is intended that the student will extend these by formulating equations and adding the conditions under which the reactions take place. The value of these charts in revision is an added advantage. Up-to-date manufacturing processes are given and notes on the uses of the more important chemicals are included. A list of ionization energies and a short bibliography conclude the book.

In conclusion, the author would like to express his appreciation to the Publishers and others who have given their valuable advice and assistance in the preparation of the present edition.

Reading, 1976   T.F.C.

CONTENTS

1 Science and Chemistry

Science is an activity concerned with finding out as much as possible about the nature of matter. In all branches of science we investigate:

(a) what materials exist in Nature;
(b) what happens, or what can be made to happen to these materials. In addition, we seek to understand why these materials do exist, and why these happenings should take place.

The biological sciences—botany, zoology, etc.—make living things the subject of their study, while the physical sciences—physics, chemistry, etc.—investigate the inanimate world.

Chemistry is the study of the nature of matter, and the ways in which materials may be produced from, or converted into other materials. Chemistry investigates:

1 *The composition of matter*, that is the identification of chemicals present in a given material, and measurement of the proportions in which they are present. These concerns of chemistry are termed qualitative analysis and quantitative analysis respectively.
2 *The properties of materials*. Experimental chemistry examines the properties and reactions of matter, and discovers methods of interconverting materials. Such experimental data are often classified under the familiar headings of preparation, properties and reactions of a substance.

Physical chemistry in particular, deals with the more quantitative aspects of experimental chemistry, and is concerned with questions such as:
the yield of a reaction,
the speed of a reaction,
the effect of a change in the external factors such as temperature or pressure on a reaction,
the heat change accompanying a reaction,
the physical properties of the substances taking part in, or produced by the reaction.

One could say that the facts of experimental chemistry are complemented by the figures of physical chemistry.

5

3 *The reason for chemical behaviour.* After the experimental data have been classified, intriguing patterns of chemical behaviour become apparent. To explain these relationships, ideas—or theories—are advanced. From these theories, the behaviour of a substance under new conditions should be predictable, so that further experimentation can be made in order to test the usefulness of each theory. This is theoretical chemistry, which evolves through imagination and insight, and, in consequence, may be less readily understood. In order to express theoretical ideas more clearly, new terminology is employed, and frequent use is made of models and analogies.

The Scientific Method

The progress of chemistry depends on advances in both theoretical and experimental chemistry. New experimental evidence often calls for a re-shaping of old ideas, while the latest theories frequently suggest new avenues of experimentation. This method—the scientific method—also forms the basis of the organization of our own laboratory work, under the three headings:

(a) *Experiment:* A full description of the experimental procedure is recorded.
(b) *Observations:* Everything that was found to take place during the experiment is carefully and accurately described.
(c) *Conclusion:* The results of the experiment are interpreted, as far as our knowledge will allow.

Very often, an experiment brings yet more questions to our notice—why was the yield so low?—why did the solution change colour?—what caused the reaction mixture to heat up?—and so on. Advances in chemistry come about because people ask questions such as these.

In studying chemistry, the student learns to:

(a) become skilled in the technique of experimentation and analysis,
(b) record accurately and honestly all his observations,
(c) examine critically both his own work and the results of others,
(d) seek a logical explanation of his observations.

2 Elements and Compounds

An *element* may be defined as a substance which cannot, by chemical means, be broken down into any simpler substances. Two or more elements may combine chemically to form a compound, in which the elements are present in fixed proportions by weight. Compounds may be distinguished from mixtures, since the properties of the individual elements are no longer apparent in a compound, while they are displayed in a mixture.

Just over one hundred elements are known, and in order to list them according to their chief chemical properties, the arrangement shown in Table 1 is used. This is known as the *extended* or *long form* of the periodic classification of elements. For convenience, the symbols for the elements are given in the table, but the names of some of the less familiar elements (up to atomic number 34) are shown below Table 1.

The periodic classification originated between 1860 and 1870, following the work of Newlands, Lothar Meyer and Mendeleeff, who, among others, listed the elements in order of increasing atomic weight. They recognized that a given element and the one seven

TABLE 1 *Modern form of the periodic classification*

Period	s-block			d-block								p-block						
	I_A	II_A	III_A	IV_A	V_A	VI_A	VII_A	VIII			I_B	II_B	III_B	IV_B	V_B	VI_B	VII_B	O
1																$_1$H				$_2$He
2	$_3$Li	$_4$Be												$_5$B	$_6$C	$_7$N	$_8$O	$_9$F	$_{10}$Ne
3	$_{11}$Na	$_{12}$Mg												$_{13}$Al	$_{14}$Si	$_{15}$P	$_{16}$S	$_{17}$Cl	$_{18}$Ar
4	$_{19}$K	$_{20}$Ca	$_{21}$Sc	$_{22}$Ti	$_{23}$V	$_{24}$Cr	$_{25}$Mn	$_{26}$Fe	$_{27}$Co	$_{28}$Ni	$_{29}$Cu	$_{30}$Zn	$_{31}$Ga	$_{32}$Ge	$_{33}$As	$_{34}$Se	$_{35}$Br	$_{36}$Kr
5	$_{37}$Rb	$_{38}$Sr	$_{39}$Y	$_{40}$Zr	$_{41}$Nb	$_{42}$Mo	$_{43}$Tc	$_{44}$Ru	$_{45}$Rh	$_{46}$Pd	$_{47}$Ag	$_{48}$Cd	$_{49}$In	$_{50}$Sn	$_{51}$Sb	$_{52}$Te	$_{53}$I	$_{54}$Xe
6	$_{55}$Cs	$_{56}$Ba	$_{57}$La*	$_{72}$Hf	$_{73}$Ta	$_{74}$W	$_{75}$Re	$_{76}$Os	$_{77}$Ir	$_{78}$Pt	$_{79}$Au	$_{80}$Hg	$_{81}$Tl	$_{82}$Pb	$_{83}$Bi	$_{84}$Po	$_{85}$At	$_{86}$Rn
7	$_{87}$Fr	$_{88}$Ra	$_{89}$Ac†															

Transition metals

* The series of elements with atomic numbers between 57 and 71 are known as the *lanthanides*.
† The series of elements with atomic numbers between 89 and 103 are known as the *actinides*.

Be beryllium B boron Ar argon Sc scandium Ti titanium V vanadium
Ga gallium Ge germanium As arsenic Se selenium

7

places beyond it in the list showed similar properties. This periodic repetition of similar chemical behaviour allowed the list to be folded back on itself to form a table containing seven columns (the noble gases were unknown at that time) containing chemically similar elements.

In the modern table, the elements are listed according to their *atomic numbers* (this number is equal to the number of protons—and hence the number of electrons—present in one atom of the element). The extended table includes the noble gases, and allows the similarities between the elements to be displayed more effectively.

Each vertical column in the Periodic Table, numbered using Roman numerals, is called a *group*. Each group is divided into two subgroups; the 'A' subgroups lie on the left of the Table, and the 'B' subgroups on the right. Group VIII is not subdivided, and it contains three columns of elements. Family names are sometimes given to certain groups of elements; for example the elements in Group IA are known as the alkali metals, while the metals in Group IIA (with the exception of beryllium) are the alkaline earths. Group 0 contains the noble gases, Group VIIB, the halogens, while the elements in Group VIB are sometimes referred to as the chalcogens.

The horizontal rows of elements are called *periods*. The first (very short) period contains hydrogen and helium only. The position of hydrogen in the Periodic Table is exceptional, since it displays properties which resemble those of the halogens, while in other instances it reacts in a similar way to the alkali metals. Thus it cannot be assigned to any one particular group.

The second and third periods each contains eight elements, while an expansion takes place in Periods 4 and 5 due to the inclusion of the transition elements. The lanthanides in Period 6 and the actinides in Period 7 are responsible for a further increase in the number of elements in each of these later periods. Bismuth, atomic number 83, is the last stable element in the Periodic Table; elements with an atomic number greater than this are radioactive, undergoing a series of nuclear changes to produce, eventually, a stable isotope.

A useful, broader type of classification may be made by combining a number of groups in the Periodic Table into sections, according to the chief chemical properties displayed by their elements.

Section 1: Metals

These are the elements (including the transition elements) which lie to the left of, and below, the stepwise dividing line running from boron to astatine (At). About 70 elements are metallic. Metals show the following typical properties:

8

(a) Lustrous appearance, malleable, ductile, high tensile strength, etc.
(b) Conduct heat and electricity.
(c) Form basic oxides.
(d) Form positive ions (cations) by losing electrons from the metal atom. Hence, metals are termed *electropositive*.

Section 2: Non-Metals

These are the elements on the right of, and above, the stepwise dividing line shown in Table 1. About 22 elements are non-metallic, and they display such typical properties as:

(a) Variable in appearance, not generally malleable or ductile.
(b) Non-conductors of heat and electricity.
(c) Form acidic oxides.
(d) Form negatively charged ions (anions) due to the gain of electrons by the non-metal atom. Hence these elements are termed *electronegative*.

Not all of these properties are displayed by every metal or non-metal, while elements lying close to the dividing line often show properties associated with both sections.

Within each of these two sections, further subdivisions can be recognized.

Section 1

The s-block metals. These are the metals of Groups IA and IIA. All these metals form colourless ions which carry a fixed charge (equal to the number of the group in which they appear), and they react readily with dilute acids or water, liberating hydrogen.

The d-block elements This section includes the transition metals (which show general properties such as forming coloured ions, exhibiting variable valency and participating in complex formation) and zinc, cadmium and mercury. However, in the case of the last three elements, d-electrons are not used in bonding so that the chemistry of these elements is more conveniently studied with the elements of the next section.

The 'B' metals These are the elements contained in the section following the transition metals and preceding the stepwise dividing line. Some of these elements resemble the more reactive metals, while

9

those lying close to the line dividing Sections 1 and 2 show varying degrees of non-metallic character.

The lanthanides   These are the elements having atomic numbers between 57 and 71. They are reactive metals showing some features typical of transition metals. The outstanding property of these elements is their very close similarity in chemical behaviour.

Section 2—The 'p-Block' Elements

The noble gases   These are the six elements contained in Group 0 of the Periodic Table. They are the least reactive of the elements— only a few compounds of xenon and krypton have yet been prepared. They are monatomic gases (i.e. they exist as single atoms) with low melting and boiling points.

The typical non-metals   These are the remaining elements in Section 2. They show the typical non-metallic properties outlined earlier.

Compounds

A compound is a chemical combination of at least two elements. Qualitative analysis identifies the elements present in a compound while quantitative analysis determines the relative proportions of the constituent elements. From these data, the empirical formula (i.e., the simplest formula that agrees with the results) for the compound may be deduced.

As an example, consider an anhydrous white salt containing 20·1 per cent magnesium, 26·6 per cent sulphur and 53·3 per cent oxygen. Dividing these percentages by the relative atomic masses gives the *atomic proportions* in which these elements are present in the compound:

$$Mg = \frac{20\cdot1}{24\cdot3} = 0\cdot827$$

$$S = \frac{26\cdot6}{32\cdot1} = 0\cdot828$$

$$O = \frac{53\cdot3}{16\cdot0} = 3\cdot331$$

These results are now divided throughout by the smallest number, so that the ratios of the numbers of atoms in the compound is found:

$$Mg = \frac{0\cdot827}{0\cdot827} = 1\cdot0$$

$$S = \frac{0\cdot828}{0\cdot827} = 1\cdot0$$

$$O = \frac{3\cdot331}{0\cdot827} = 4\cdot0$$

From these results it follows that the empirical formula is $MgSO_4$.

Atoms, Relative Atomic Masses and Moles

Although modern physics has shown an atom to have a complex structure, Dalton's idea of an atom as the smallest particle of matter that can exist, is an extremely useful concept in chemistry. In chemical reactions, the complex structure we call an atom functions as a single unit, whereas in a nuclear reaction, the separate parts of the atom are involved and the unit is changed.

The relative atomic mass of an element is the ratio of the weight of one atom of an element to that of one atom of a reference element. Modern methods of determining relative atomic masses involve the use of a mass spectrograph, and the current definition of relative atomic mass is related to this technique.

Relative atomic mass =
$$\frac{\text{weight of one atom of the element}}{\text{weight of one atom of carbon 12 isotope}} \times 12$$

(The multiplying factor 12 is introduced to avoid atomic masses of less than unity.)

Although relative atomic masses have no units, the use of g atoms (i.e. the atomic mass expressed in grams) or g molecules (the molecular mass expressed in grams) is widespread. Use of the term 'molecular weight' has been criticized on the grounds that ionic substances such as sodium chloride cannot have a molecular weight in the strict sense. Consequently, for the expression of quantities of material, the idea of a *mole* has been introduced. Exactly 12 g of the isotope carbon 12 contains a huge, but finite number of atoms. This number, known as *Avogadro's constant* (symbol L), has been found to be $6\cdot02 \times 10^{23}$. Thus, a mole is defined as the amount of a substance which contains Avogadro's constant of chemical units.

For an element, one mole is the relative atomic mass in grams; for a compound, one mole is contained in the formula weight in grams. As examples,

11

1 mole of argon (Ar) is equal to 40 g
1 mole of hydrogen (H_2) weighs 2 g
1 mole of sodium hydroxide (NaOH) weighs $23 + 16 + 1 = 40$ g

The idea of a mole representing a fixed number of units is a useful concept, but *it is important to realize exactly what the chemical units are.*

For example, dissolving 1 mole of barium chloride crystals in water, produces a solution containing 1 mole of barium ions, but 2 moles of chloride ions.

Valency and Oxidation Numbers

Valency is a term used to indicate the combining power of an element. This, in turn, is derived from a consideration of the number of atoms with which one atom of a given element appears to be combined. It is not always easy to see, from the formula for a compound, just how the atoms are combined, and thus the value of the valency of an element is uncertain. To avoid these difficulties, the concept of *oxidation number* (or *state*) is used as a means of indicating the combining power of an element as judged from the formula for a particular compound.

Assignment of Oxidation Numbers

1 The oxidation number of an element
   (a) in combination with itself, is zero (e.g., H_2, P_4, S_8, etc.);
   (b) in combination with other elements, is either positive or negative, and, occasionally, zero.
2 The s-block metals have positive oxidation numbers equal to their group number. Thus, the oxidation number of potassium is $+1$, and the oxidation number of calcium is $+2$ in all their compounds.
3 Transition metals have varying oxidation numbers. For example, manganese has the oxidation numbers $+2$ in $MnCl_2$, $+4$ in MnO_2 and $+7$ in $KMnO_4$.
4 Fluorine has an oxidation number of -1, oxygen -2 (except in combination with fluorine, when it is $+2$, and in peroxides $(-O-O-)$, when the value is -1). Hydrogen usually shows an oxidation state of $+1$, but it is -1 in hydrides such as LiH and CaH_2.
5 The sign of the oxidation number for the other non-metals is deduced from the relative electronegativity of the elements. The most electronegative element has a negative oxidation number.
6 The oxidation number of an ion is equal to the charge carried by

12

the ion, and the algebraic sum of the oxidation numbers of the atoms forming an ion, is equal to the charge on the ion. The algebraic sum of the oxidation numbers of the atoms forming a neutral molecule is zero.

As an example, the oxidation number of the sulphate ion SO_4^{2-} is -2, while the oxidation number of oxygen is -2. The total oxidation number for four oxygens is -8. Since the algebraic sum of the oxidation numbers is -2, the oxidation number of sulphur must be $+6$. Also, in nitric acid, HNO_3, oxygen is the most electronegative element, while hydrogen is the least electronegative. Hence, the total oxidation number for three oxygens is -6, and the oxidation number of hydrogen is $+1$, which means that the oxidation number of nitrogen is $+5$, since the sum of these is zero.

7 When the oxidation state of an element in a compound is not obvious, it can usually be calculated by applying the above rules. As an example, we calculate the oxidation numbers in $KClO_4$:

$$\begin{aligned} \text{Oxidation number of potassium} &= +1 \\ \text{Oxidation number for four oxygens} &= -8 \\ \hline \text{Total} &= -7 \end{aligned}$$

To balance, the oxidation state of chlorine in this compound must be $+7$.

As a second example, the oxidation number of iron in $K_3[Fe(CN)_6]$ can be deduced as follows:

$$\text{Total oxidation number of the ion } [Fe(CN)_6]^{3-} = -3$$

But,

$$\text{total oxidation number for six cyanide ions} = -6$$

Hence,

$$\text{oxidation number of iron} = +3.$$

Oxidation numbers are useful as a means of defining oxidation and reduction, and in compiling oxidation state charts.

Naming Compounds

The name given to a compound should be unambiguous, and should give a clear indication of the formula of the compound. To this end, the International Union of Pure and Applied Chemistry have drawn up a system of nomenclature, and some of the important rules are as follows:

1 Both in the chemical formula, and in the English name of the compound, the most electropositive element is given first; $CaCl_2$,

13

and not Cl_2Ca. Many compounds contain more than one non-metallic element, and for these cases, the element appearing first in the following series is written first:

> B,
> Si, C
> Sb, As, P, N.
> H
> Te, Se, S, O
> I, Br, Cl, F.

Consequently, the formula SiC is correct, and the compound is called silicon carbide, not carbon silicide. The formulae NH_3 and H_2S are correct, but F_2O should be written OF_2 and called oxygen fluoride.

2 In a compound consisting of two elements only, the name of the second element is modified to terminate in -ide; for example, NaCl, sodium chloride.

3 The numbers of atoms present in the formula for a compound may be indicated by the following prefixes to the name:

$\frac{1}{2}$	hemi	3	tri	7	hepta
1	mono	4	tetra	8	octa
	(often omitted)	5	penta	9	ennea
2	di	6	hexa	10	deca

Examples:

NO	nitrogen monoxide (or nitrogen oxide)
CO_2	carbon dioxide
NF_3	nitrogen trifluoride
N_2O	dinitrogen oxide
NO_2	nitrogen dioxide
N_2O_4	dinitrogen tetra-oxide

4 Salts: these are named according to the *Stock* notation, in which the name of the salt, and the formula (if necessary) include the oxidation state of the metal, using Roman numerals. For example:

$Cu^{II}SO_4$	copper (II) sulphate
$FeCl_3$	iron (III) chloride

Acid salts are named according to the number of hydrogen atoms present in the salt;

$NaHCO_3$	sodium hydrogen carbonate (not bicarbonate)
NaH_2PO_4	sodium dihydrogen phosphate (V)

14

5 For salts of oxyacids, the name ending is modified to end in -ate, with the oxidation state of the non-metal being given (in brackets).

NaClO	sodium chlorate (I)
$NaClO_2$	sodium chlorate (III)
$NaClO_3$	sodium chlorate (V)
$NaClO_4$	sodium chlorate (VII)

The oxidation numbers of chlorine are $+1$, $+3$, $+5$ and $+7$ respectively.

6 Some well-known common names (water, ammonia) are retained.

3 Atomic Structure and Electronic Configuration

The existence of many particles smaller than an atom has been recognized. Of these, three particles—neutrons, protons and electrons—are of special importance in chemistry. Their chief properties are listed in Table 2.

TABLE 2 *Properties of the three fundamental particles in an atom*

Particle	Mass (H = 1)	Charge	Location
Neutron	1	0	In nucleus
Proton	1	+1	In nucleus
Electron	1/1846	−1	Surrounding nucleus

The radius of an atom is of the order of 10^{-10} m or 10^{-1} nm. In an atom, the neutrons and protons are closely associated to form a central core or *nucleus*, while the electrons occupy the volume between the nucleus and the periphery of the atom. Protons carry a positive charge equal in magnitude (1.6×10^{-19} C) to the negative charge carried by the electron. The neutron is uncharged.

The mass of the electron is usually neglected in computing the relative atomic mass of an atom, which is therefore regarded as the sum of the masses of the protons and neutrons (i.e. the sum of the numbers of these particles) contained in the atom. Therefore, for an atom of relative atomic mass A,

$$A = p + n$$

(where p and n represent the numbers of protons and neutrons, respectively, present in that atom).

The atomic number Z is equal to the number of protons (or electrons) in the atom, so that

$$Z = p \quad (= e)$$

16

The difference between the relative atomic mass and the atomic number represents the number of neutrons in the atom:

$$A - Z = p + n - p = n$$

Atoms of the same element must have the same number of protons in their nuclei, but the number of neutrons may vary, thus giving rise to atoms of the same chemical element having different relative atomic masses. Such atoms are said to be *isotopes*. For example, about 75 per cent of chlorine atoms have 17 protons and 18 neutrons in their nuclei, making the relative atomic mass 35. Of the remainder, almost all have 17 protons and 20 neutrons, giving a relative atomic mass of 37. Thus the average relative atomic mass for a large assembly of chlorine atoms will be close to the value calculated as follows:

$$\text{Average relative atomic mass} = \frac{35 \times 75}{100} + \frac{37 \times 25}{100} = 35 \cdot 5$$

Sometimes (e.g., in discussing nuclear reactions) it is necessary to specify exactly which isotope is meant when writing the symbol for an element. This is done by writing the relative atomic mass as a superscript, and the atomic number as a subscript, in front of the symbol for the element. For example, $^{35}_{17}Cl$ and $^{37}_{17}Cl$ indicate the light and heavy isotopes of chlorine respectively.

The following properties of electrical charge are of importance in determining the electronic configuration of an atom:

1 Equal quantities of positive and negative charge placed at the same point in space, neutralize the effects of each other.
2 A force exists between two charges (of magnitude q_1 and q_2) placed distance d apart, in a medium of dielectric constant k, which is given by the equation:

$$\text{Force} = \frac{q_1 \times q_2}{k \times d^2}$$

This is *Coulomb's inverse square law*. The force is attractive when the charges are opposite in sign and repulsive when they are of the same sign.
3 A moving charge produces a magnetic effect.

The Electronic Configuration of the Elements

Before atoms combine, they must first come into close association with each other. This means that the outermost electrons (i.e., those furthest from the nucleus) are involved in the process of combination. Since some atoms combine readily (e.g., hydrogen and chlorine) while others do not (e.g., the noble gases) it would appear that the 'make-up' or configuration of the outermost electron screen is an important factor in chemical combination.

To determine the electronic configuration of an atom, we need to know:

(a) the number of electrons present in the atom,
(b) the way in which these electrons are arranged in the space surrounding the nucleus.

The answers to these problems lie in the use of ionization energies, the study of atomic spectra, and in the development of the quantum theory.

Ionization Energy

This is the minimum amount of energy required to remove the most loosely bound electron from 1 mole of gaseous atoms of an element. The process which corresponds to the equation

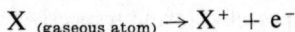

$$X_{\text{(gaseous atom)}} \rightarrow X^+ + e^-$$

is called the first ionization energy of X, and is expressed in joules per mole of atoms of X.

A further amount of energy—the second ionization energy, is needed to withdraw a second electron from X^+, and so on, until the n-th ionization energy removes the n-th (the final) electron. Ionization energy data are available for most elements, and, using the data in the Appendix at the end of the book, a graph of the successive ionization energies for a given element may be plotted. Figure 1 shows the graph for neon, in which the logarithm of the ionization energy has been plotted in order to keep the scale to a convenient size.

Deductions from the graph (Figure 1)

1 Only ten successive ionization energies can be measured; hence a neon atom contains ten electrons, and, therefore, ten protons, in agreement with the atomic number of neon.
2 The ten electrons are arranged in two groups, one containing eight electrons, and the other, two (giving neon the configuration 2.8).

3 The group of two electrons lies closer to the nucleus, as the electrons in this group are more difficult to remove.
4 The group of eight electrons lies further from the nucleus, corresponding to the lower ionization energies needed for the removal of these electrons.

Note: A general upward trend in successive ionization energies is expected, since the removal of an electron from an atom means that:

   (a) there are more protons effectively attracting the remaining electrons in the ion;
   (b) the ion is smaller in size than the parent atom.

Both these factors make for a steady increase in the successive ionization energies for electrons in the same group. Breaks in this pattern of steady increase indicate the grouping of the electrons.

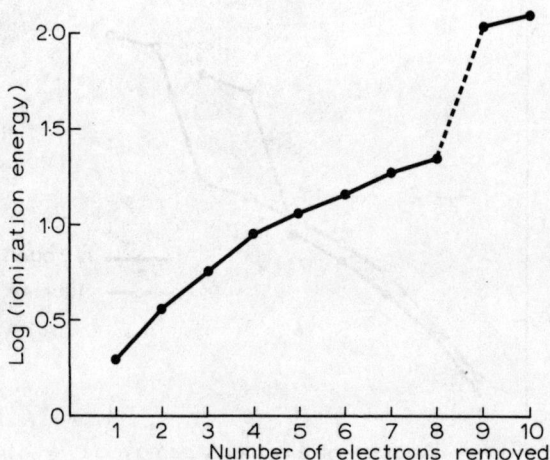

Fig. 1

The information so far deduced from this type of graph is displayed diagrammatically in Figure 2. Note that the electrons in the first electron energy level (i.e., those nearest to the nucleus) are often said to be in a 'lower energy level', yet it is clear from the following diagram that they require a larger amount of energy in order to withdraw them from this level.

19

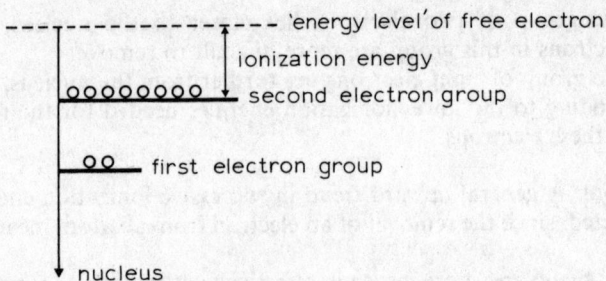

Fig. 2 Electron energy levels for neon

Figure 3 shows the corresponding graphs for nitrogen and fluorine, and we can recognize electronic groupings of 2 and 5 for nitrogen, and 2 and 7 for fluorine respectively. However, when the successive ionization energies for the outermost group of electrons (i.e., the first five ionization energies for nitrogen, or the first seven for

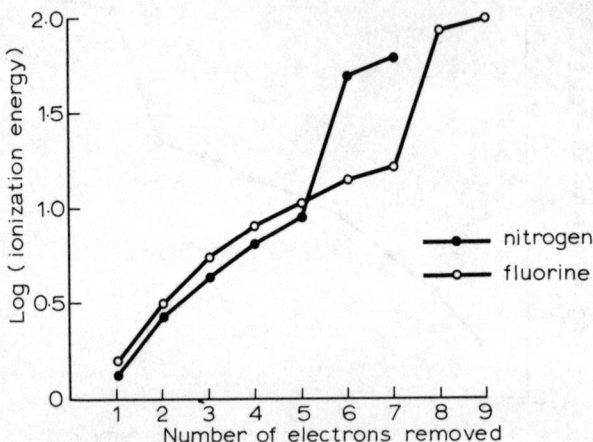

Fig. 3 Successive ionization energies for nitrogen and fluorine

fluorine) are plotted on a large scale, another slight but significant break in the pattern of steady increase is seen. Figure 4 shows the graph for fluorine, where an approximately steady increase takes place until five electrons have been removed. A break in the continuity is seen corresponding to the removal of the sixth electron. Thus it is concluded that the outer group of seven electrons is subdivided

20

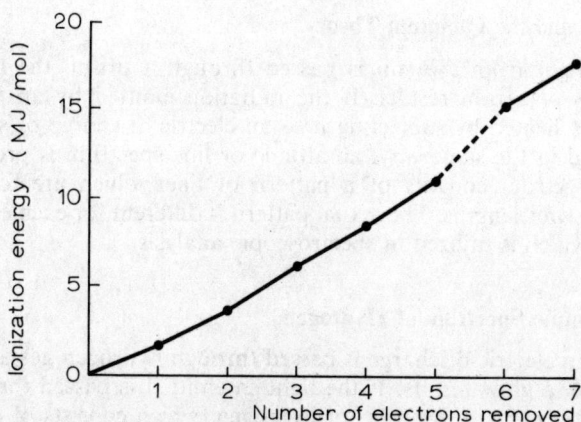

Fig. 4 The first seven ionization energies of fluorine

into two sub-levels, one containing two electrons (which are more difficult to ionize), with five electrons in the other (more readily ionizable) sub-level.

By plotting this large-scale type of graph, using the data given in the Appendix, the electronic configurations for the elements shown in Table 3 can be deduced.

TABLE 3 *The electronic configurations of the elements as deduced from ionization energies*

Element	Atomic number	Electronic configuration (*near nucleus*)		(*further from nucleus*)
H	1	1	–	–
He	2	2	–	–
Li	3	2	1	–
Be	4	2	2	–
B	5	2	2·1	–
C	6	2	2·2	–
N	7	2	2·3	–
O	8	2	2·4	–
F	9	2	2·5	–
Ne	10	2	2·6	–
Na	11	2	2·6	1
Mg	12	2	2·6	2
Al	13	2	2·6	2·1
Si	14	2	2·6	2·2
P	15	2	2·6	2·3
S	16	2	2·6	2·4
Cl	17	2	2·6	2·5
Ar	18	2	2·6	2·6

21

Spectra and the Quantum Theory

When light from the sun is passed through a prism, the familiar rainbow spectrum results. If the radiation emitted by an element which is heated by subjecting it to an electric discharge or spark is analysed in the same way, an atomic or line spectrum is produced. This spectrum consists of a pattern of lines which are found at various wavelengths. The actual pattern is different for each element, a fact which is utilized in spectroscopic analysis.

The Atomic Spectrum of Hydrogen

When an electric discharge is passed through hydrogen gas at a low pressure, a glow results. If the light so emitted is passed through a prism, the atomic spectrum of hydrogen, which consists of a series of sharp lines as recorded on a photographic plate is produced. Figure 5 shows the first four lines in the visible region of the spectrum.

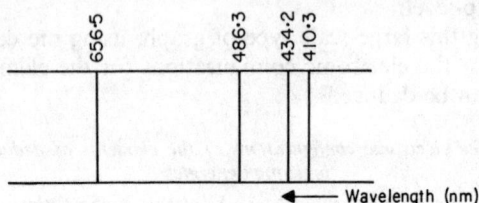

Fig. 5 First four lines in the Balmer series

In 1885, Balmer showed that the wavelengths of these lines could be related by an equation of the form

$$\frac{1}{\lambda} = \text{constant} \left(\frac{1}{2^2} - \frac{1}{n^2} \right)$$

where n is a whole number greater than 2. When $n = 3$, $\lambda = 656 \cdot 475$ nm, and when $n = 4$, $\lambda = 486,273$, etc. The constant is the *Rydberg* constant (R) which has a value $1 \cdot 097 \times 10^7 \text{ m}^{-1}$.

Further work showed that other series of lines were present in the atomic spectrum of hydrogen, the wavelengths of the lines within each series being related by an equation

$$\frac{1}{\lambda} = R \left(\frac{1}{n_1^2} - \frac{1}{n_2^2} \right) \quad \text{(where } n_2 > n_1 \text{)}$$

22

When $n_1 = 1$, the series is known as the Lyman series and it is located in the ultraviolet region of the electromagnetic spectrum, while n_1 has a value of 3 in the Paschen series which is found in the infrared.

Interpretation of Atomic Spectra

An explanation of the atomic spectra of hydrogen was given by Bohr in 1914. This theory proposed that the electron in a hydrogen atom was circling the nucleus, the electrostatic attraction between the nucleus and the electron balancing the tendency of the electron to fly off at a tangent.

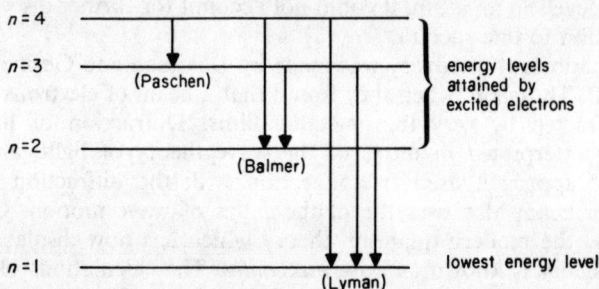

Fig. 6 Energy levels

As a result, the electron has

(a) kinetic energy of motion,
(b) potential energy resulting from the electrostatic attraction.

The sum of the kinetic and potential energies is the total energy E of the electron.

In order to explain certain features connected with the radiation of heat energy by hot bodies, Planck, in 1900, introduced the *quantum theory*, which put forward the idea that radiant energy could be absorbed or emitted in fixed amounts, called *quanta*. (This is similar to money changing hands in fixed units, the quantum being the smallest coin of the realm.) The energy of one quantum of radiation is related to the wavelength of the radiation by the equation

$$E = \frac{hc}{\lambda}$$

23

where h is Planck's constant, c is the velocity of light, and λ is the wavelength of the radiation.

In the case of the hydrogen atom, the quantum theory would require the total energy of the electron to be restricted to a fixed set of values, so that an electron, excited by the electric discharge, would reach a higher energy level (as shown in Figure 6). When the electron falls back to its original level, radiation corresponding to a spectral line is emitted.

The Modern Quantum Theory

Although the Bohr theory was successful in explaining the broad features of line spectra, and in establishing the idea of electronic energy levels in an atom, it could not account for further discoveries in relation to line spectra.

An important discovery was made by Davisson and Germer, and by G. P. Thomson, when they found that a beam of electrons could be diffracted by very thin metallic films. Diffraction of light is usually interpreted in terms of the wave theory of light, and the parallel approach used in connection with the diffraction of an electron beam also uses the mathematics of wave motion. Consequently, the modern quantum theory which has now displaced the Bohr theory is known as *wave mechanics*. The calculations of wave mechanics are extremely complicated, but the results play an important part in giving us a picture of the electronic configuration of an atom.

Wave Mechanics and Electronic Configuration

Wave mechanics shows that:

1 The total energy that an electron in an atom can possess is limited to a restricted set of values called *atomic energy levels*.
2 These energy levels are described by a principal quantum number n. For the first level, $n = 1$, and electrons in this energy level have the highest ionization energies (and therefore lie closest to the nucleus). Further successive energy levels are those for which $n = 2$, 3 and so on.
3 Each principal energy level may be split into a number of sub-levels, according to the values of the *subsidiary quantum number l*. The variable l can have the value zero, and all positive whole number values up to $n - 1$. Thus, when $n = 1$, $l = 0$, which means that the first level is not split. When $n = 2$, l can be 0 or 1, so that the second principal energy level is subdivided into two.

In fact, the number of sub-levels is equal to the value of the principal quantum number n.

4 Two further quantum numbers, m and s, are involved. The *magnetic quantum number m* can take the values zero and all positive and negative whole numbers between $+l$ and $-l$. Thus, when $l = 2$, m can be $+2$, $+1$, 0, -1 and -2, that is it can take five different values. It can easily be seen that there are $2l + 1$ separate values of m for any given value of l.

The *spin quantum number s* has two values, $+\frac{1}{2}$ and $-\frac{1}{2}$.

The significance of the number of distinct m values for each value of l, and of the spin quantum number, will be dealt with later.

5 Each sub-level in this system is described by a symbol depending on the n and l values. Any sub-level for which $l = 0$ is designated by the symbol s (not to be confused with the spin quantum number s). When $l = 1$, the symbol is p; for $l = 2, 3, 4$, etc. the symbols are d, f, g (and so on alphabetically). The value of n is placed in front of each symbol. Thus, 1s describes the energy level for which $n = 1$ and $l = 0$, while 3d corresponds to $n = 3, l = 2$. The diagram given in Figure 7 illustrates this.

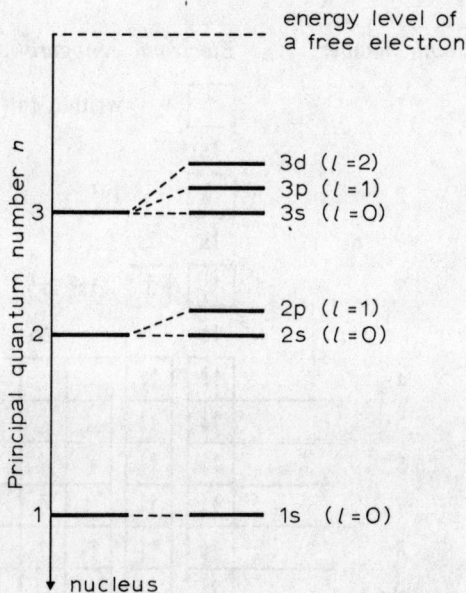

Fig. 7 Atomic energy levels

25

Significance of the m and s quantum numbers Since there are $2l + 1$ values of m for a given value of l, it follows that, in the 2p-level (or in fact, any p-level), there are three values of m ($+1$, 0 and -1). Each value of m is said to correspond to an *atomic orbital*, so that there are, in fact three p-orbitals (labelled p_x, p_y and p_z) for every value of n greater than 1. Similarly, there is only a single s-orbital for each value of n, and five d-orbitals (when $n = 3$ or more).

Three important rules follow:

1 *Pauli's exclusion principle* which states that 'No two electrons in an atom can be described by the same set of four quantum numbers'. This leads to the conclusion that each orbital can accept a maximum of two electrons, one with an s value of $+\frac{1}{2}$, the other $-\frac{1}{2}$. A full description of the electronic configurations of the elements can now be made, and this is usually done by drawing a series of boxes, labelled with the orbital symbol, and placing in these boxes one or two arrows (one pointing upwards to show $s = +\frac{1}{2}$, the other downwards for $s = -\frac{1}{2}$) to represent electrons occupying the orbitals.

Element	Atomic number	Electronic configuration

Element	Atomic number	1s	2s	2p			Electronic configuration
H	1	↑					written $1s^1$
He	2	↑↓					$1s^2$
Li	3	↑↓	↑				$1s^2\,2s^1$
Be	4	↑↓	↑↓				$1s^2\,2s^2$
B	5	↑↓	↑↓	↑			$1s^2\,2s^2\,2p^1$
C	6	↑↓	↑↓	↑	↑		$1s^2\,2s^2\,2p^2$
N	7	↑↓	↑↓	↑	↑	↑	$1s^2\,2s^2\,2p^3$
O	8	↑↓	↑↓	↑↓	↑	↑	$1s^2\,2s^2\,2p^4$
F	9	↑↓	↑↓	↑↓	↑↓	↑	$1s^2\,2s^2\,2p^5$
Ne	10	↑↓	↑↓	↑↓	↑↓	↑↓	$1s^2\,2s^2\,2p^6$

2 *Hund's rule.* Electrons are arranged in a set of orbitals of equal energy (i.e., $2p_x$, $2p_y$, $2p_z$) with parallel spins as far as possible. Thus, the sixth electron in carbon enters an unoccupied p-orbital with spin in the same sense as the p-electron already present. Nitrogen has one electron (with parallel spin) in each p-orbital.

3 Electrons are accommodated in the lowest energy level available. The order of increasing orbital energies is given by the *aufbau*

TABLE 4 *Electronic configuration of the elements*
(atomic numbers 1–36)

Element	Atomic number	Electronic configuration
H	1	$1s^1$
He	2	$1s^2$
Li	3	$1s^2\ 2s^1$
Be	4	$1s^2\ 2s^2$
B	5	$1s^2\ 2s^2\ 2p^1$
C	6	$1s^2\ 2s^2\ 2p^2$
N	7	$1s^2\ 2s^2\ 2p^3$
O	8	$1s^2\ 2s^2\ 2p^4$
F	9	$1s^2\ 2s^2\ 2p^5$
Ne	10	$1s^2\ 2s^2\ 2p^6$
Na	11	$1s^2\ 2s^2\ 2p^6\ 3s^1$
Mg	12	$1s^2\ 2s^2\ 2p^6\ 3s^2$
Al	13	$1s^2\ 2s^2\ 2p^6\ 3s^2\ 3p^1$
Si	14	$1s^2\ 2s^2\ 2p^6\ 3s^2\ 3p^2$
P	15	$1s^2\ 2s^2\ 2p^6\ 3s^2\ 3p^3$
S	16	$1s^2\ 2s^2\ 2p^6\ 3s^2\ 3p^4$
Cl	17	$1s^2\ 2s^2\ 2p^6\ 3s^2\ 3p^5$
Ar	18	$1s^2\ 2s^2\ 2p^6\ 3s^2\ 3p^6$
K	19	(—Ne—) $3s^2\ 3p^6\ 4s^1$
Ca	20	(—Ne—) $3s^2\ 3p^6\ 4s^2$
Sc	21	(—Ne—) $3s^2\ 3p^6\ 3d^1\ 4s^2$
Ti	22	(—Ne—) $3s^2\ 3p^6\ 3d^2\ 4s^2$
V	23	(—Ne—) $3s^2\ 3p^6\ 3d^3\ 4s^2$
Cr	24	(—Ne—) $3s^2\ 3p^6\ 3d^5\ 4s^1$
Mn	25	(—Ne—) $3s^2\ 3p^6\ 3d^5\ 4s^2$
Fe	26	(—Ne—) $3s^2\ 3p^6\ 3d^6\ 4s^2$
Co	27	(—Ne—) $3s^2\ 3p^6\ 3d^7\ 4s^2$
Ni	28	(—Ne—) $3s^2\ 3p^6\ 3d^8\ 4s^2$
Cu	29	(—Ne—) $3s^2\ 3p^6\ 3d^{10}\ 4s^1$
Zn	30	(—Ne—) $3s^2\ 3p^6\ 3d^{10}\ 4s^2$
Ga	31	(—Ne—) $3s^2\ 3p^6\ 3d^{10}\ 4s^2\ 4p^1$
Ge	32	(—Ne—) $3s^2\ 3p^6\ 3d^{10}\ 4s^2\ 4p^2$
As	33	(—Ne—) $3s^2\ 3p^6\ 3d^{10}\ 4s^2\ 4p^3$
Se	34	(—Ne—) $3s^2\ 3p^6\ 3d^{10}\ 4s^2\ 4p^4$
Br	35	(—Ne—) $3s^2\ 3p^6\ 3d^{10}\ 4s^2\ 4p^5$
Kr	36	(—Ne—) $3s^2\ 3p^6\ 3d^{10}\ 4s^2\ 4p^6$

(or 'building up') principle. This is best illustrated diagrammatically. In this diagram, the orbitals corresponding to each principal

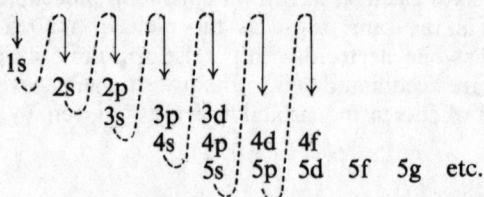

quantum number are written on a separate line, and indented one
place to the right (as one addresses a letter). The sequence of
increasing orbital energies is given by the order in which they
appear in the vertical columns, taking the left-hand column first.

Following these rules, the electronic configuration of the elements
can be deduced. Table 4 shows the electronic configurations of the
first thirty-six elements.

Special Features Relating to Electronic Configuration

1 Elements in the same group have similar electronic configurations.
  For example, carbon, silicon and germanium have the configuration $s^2 p^2$. To show that the s- and p-orbitals belong to the same
  principal quantum level, the general designation $ns^2 np^2$ is used.
2 The electrons concerned with bond formation are termed *valence*
  electrons. Valence electrons are usually those present in orbitals
  having the highest quantum number. The transition elements are
  an exception to this generalization, since the $(n - 1)$d-electrons
  and the ns-electrons are involved in bonding.
3 With scandium (atomic number 21) and the first row transition
  metals, the 4s-orbital is occupied before the 3d-orbitals. This is a
  consequence of the *aufbau* principle, which leads to the natural
  inclusion of the transition metals at this point, as part of the
  steady build-up of electronic configuration.
4 In Table 4, although the 3d-orbitals are filled after the 4 s-orbital,
  the configurations are written in the order shown. Thus the general
  rule is that the first electron (or electrons) to be removed on
  ionization are those present in the last-named orbital in the
  configuration of a given element. For example, vanadium,
  (—Ne—) $3s^2 3p^6 3d^3 4s^2$, loses the two 4s-electrons first on ionization, as can be seen from the graph (Figure 8) showing the first
  five ionization energies for vanadium.
5 The out-of-sequence configurations of copper and chromium,
  depend on the special stability which appears to be characteristic

28

of a half filled, or a completely filled, set of orbitals. With chromium, the configuration $3d^5\ 4s^1$ leaves both levels half filled, which is preferred to $3d^4\ 4s^2$. Similarly, with copper, the configuration $3d^{10}\ 4s^1$ fills the 3d-level and half fills the 4s-level.

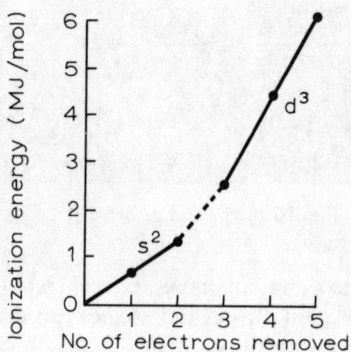

Fig. 8 Ionization energies for vanadium

Shapes and Sizes of Orbitals

Further results of wave mechanics show that:

1 An electron cannot be located precisely within the atom.
2 Atomic orbitals indicate the regions in the space surrounding the nucleus of an atom, in which an electron is most likely to be found.
3 The shape of an orbital depends on the value of the quantum number l. For an s-orbital ($l = 0$) the shape is a sphere, showing that all directions are equally likely to be inhabited by the s-electrons. The fact that a 2s-electron is, on average, at a greater distance from the nucleus than a 1s-electron, is indicated by drawing a larger sphere for the 2s-orbital.

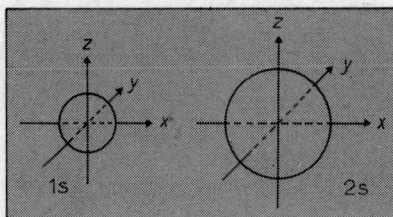

Fig. 9 1s and 2s orbitals

29

In the case of the three p-orbitals ($l = 1$), the region most likely to hold the p_x-electrons is dumbell shaped, and extends in both the positive and negative directions of the x-axis.

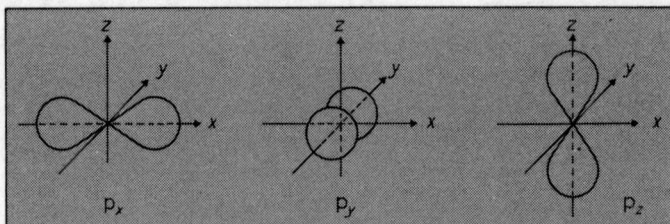

Fig. 10 p_x, p_y and p_z orbitals

p_y- and p_z-orbitals have similar shapes, concentrated along the y and z axes respectively. Again, 3p-orbital shapes are drawn larger than the 2p-orbital shapes, since the electrons in the 3p-orbitals are more likely to be found at a greater distance from the nucleus than the 2p-electrons.

The shapes of the five d-orbitals are more complicated, and the reader is referred to more advanced textbooks for the outline drawings of these orbitals.

4 Bonding

A compound is formed by the combination of at least two elements, and the linkage which unites the atoms in a compound is a bond. Bonds may be classified into four distinct types, as shown in Table 5.

TABLE 5 *The four 'ideal' types of bonding*

Type and example	Mechanism	Characteristic properties of compound
1 Ionic: Na^+Cl^-	Electrostatic forces between ions	Form hard, brittle crystals, with high melting and boiling points. Soluble in polar solvents, and conduct electricity (with decomposition) when in solution or fused
2 Covalent: CH_4, NF_3	Electron sharing to give an electron pair bond	Either: (a) discrete molecules with low melting and boiling points. Soluble in non-polar solvents, and non-conductors of electricity, or, (b) 'giant' molecules, hard, high melting point and boiling point. Insoluble in most solvents
3 Metallic: sodium, iron	Metal ions held in a crystal lattice by a large assembly of electrons	Hard and crystalline. High melting and boiling points. Usual metallic properties—high tensile strength, ductility and malleability, etc. Conduct electricity in solid state without decomposition
4 Van der Waals: liquid and solid noble gases	Van der Waals forces	Low melting and boiling points. Soluble in non-polar solvents

These four types of bonding represent ideal cases. Very many compounds are formed in which the bonding is best described as 'intermediate' in character, although one or other of the above types usually predominates.

The Ionic Bond

In this type of bonding, one or more electrons are transferred from one atom to another. This process is possible if:

(a) electrons are readily lost by one of the atoms (i.e., the element has a low ionization energy) to form a cation. The metallic elements behave in this way.

(b) electrons are readily accepted by the other atom, which forms an anion. Atoms of the more electronegative non-metals react in this way.

The process is represented by the following model, taking sodium fluoride as an example.

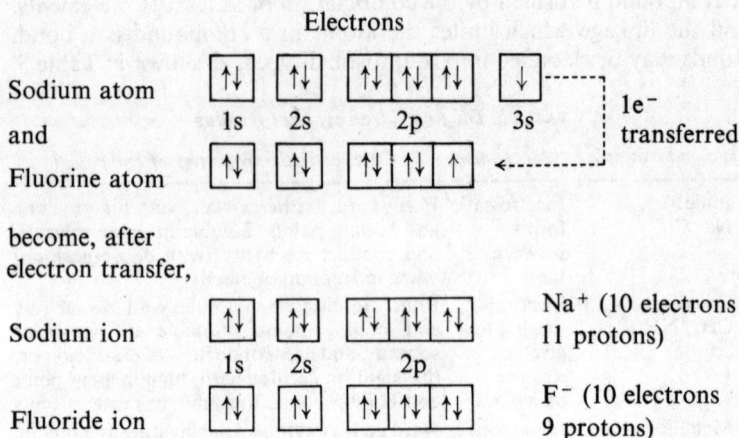

Electrons

Sodium atom   [↑↓] [↑↓] [↑↓][↑↓][↑↓] [↓] ┐
and   1s   2s   2p   3s │ 1e⁻
                                               │ transferred
Fluorine atom [↑↓] [↑↓] [↑↓][↑↓][↑] ◄──────┘

become, after electron transfer,

Sodium ion   [↑↓] [↑↓] [↑↓][↑↓][↑↓]   Na⁺ (10 electrons
               1s   2s   2p   11 protons)

Fluoride ion [↑↓] [↑↓] [↑↓][↑↓][↑↓]   F⁻ (10 electrons
                                                   9 protons)

When an ionic compound is produced, even on the smallest scale, a very large number of ions are formed. By mutual attraction and repulsion, the ions settle out into a regular arrangement called a crystal lattice. A crystal is a solid of regular shape, bounded by flat faces. The external symmetry of a crystal is due to the regular internal packing.

Ionic Crystals

1 Composed of a regular assembly (crystal lattice) of cations and anions.
2 In general, anions are larger than cations.
3 Ions are packed together as closely as possible, with the maximum number of oppositely charged ions surrounding any given ion.
4 The number of ions (of opposite charge) which are nearest to, and equidistant from a given ion, is known as the *crystal co-ordination* number of that ion.
5 For electroneutrality, the total charge carried by the cations is equal to the total charge carried by the anions in the crystal.

The relative sizes of the anions and cations affect the numbers of ions which may cluster round each other, and this governs the pattern in which the lattice builds up. The ratio, radius of cation: radius of anion, is called the *radius ratio*. Its values correspond to different typical crystal structures. Two of these, the *rock salt* and the *caesium chloride* structures, are shown in Figure 11.

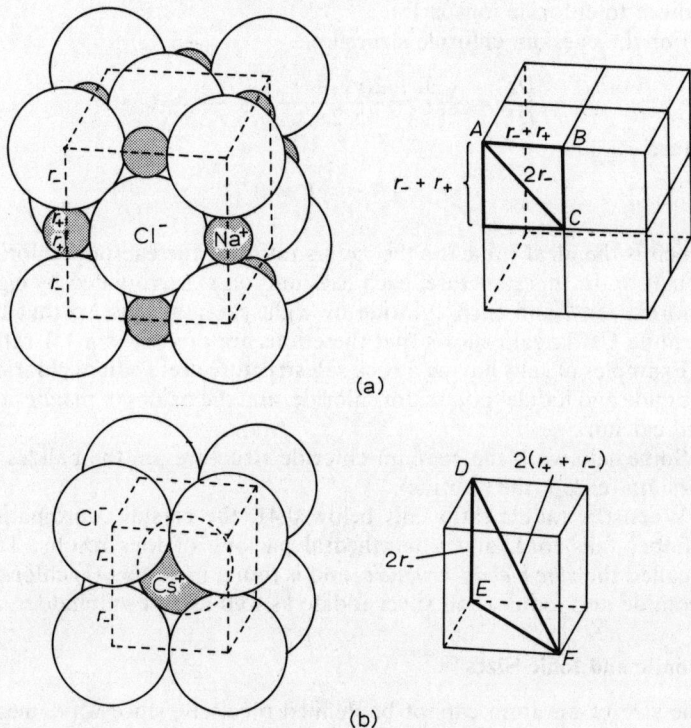

(a)

(b)

Fig. 11 Typical crystal structures. (a) Sodium chloride (rock salt) crystal. (b) Caesium chloride crystal

In the sodium chloride crystal, the ratio

$$\frac{AB}{AC} = \frac{r_+ + r_-}{2r_-} = \frac{1}{\sqrt{2}}$$

Hence,

$$\frac{r_+}{r_-} + 1 = \frac{2}{\sqrt{2}} = \sqrt{2} = 1.41$$

Thus,

$$\frac{r_+}{r_-} = 1.41 - 1 = 0.41$$

33

which is the ideal value for the radius ratio for this type of crystal structure.

Each sodium ion in the rock salt structure is surrounded by six chloride ions, and hence the crystal coordination number of sodium is 6. Similarly, the crystal coordination number of the chloride ion is 6, and the formula NaCl shows that the ratio of the numbers of sodium to chloride ions is 1:1.

For the caesium chloride structure,

$$\frac{DF}{DE} = \frac{\sqrt{3}}{1} = \frac{2(r_+ + r_-)}{2r_-} = \frac{r_+}{r_-} + 1$$

Thus,

$$\frac{r_+}{r_-} = 1\cdot73 - 1 = 0\cdot73$$

which is the ideal value for the radius ratio for the caesium chloride structure. In this structure, each caesium ion is surrounded by eight chloride ions, and each chloride by eight caesium ions, so that the formula CsCl again shows that these ions are present in a 1:1 ratio.

Examples of salts having a rock salt structure are; sodium chloride, bromide and iodide, potassium chloride, and the oxides of magnesium and calcium.

Some salts with the caesium chloride structure are the halides of caesium (except the fluoride).

When the radius ratio falls below 0·41, the crystal coordination number falls to 4, and a tetrahedral packing of ions results. This is called the *zinc blende structure*, and is found in copper (I) chloride, bromide and iodide, and silver iodide as well as zinc sulphide.

Atomic and Ionic Sizes

The size of an atom cannot be defined precisely, since wave mechanics shows that there is *no* definite boundary which may be drawn round an atom, outside which the electrons cannot be found. However, the distance between neighbouring nuclei can be determined experimentally, and a self-consistent set of values of 'atomic radii' can be drawn up. The size of an atom increases with increase in the atomic number down a group; Li 0·123 nm, Na 0·157 nm, K 0·203 nm, etc. This trend is expected as it corresponds to the addition of electrons in successive quantum levels. Across a period, the atomic size gradually decreases: Li 0·123 nm, Be 0·089 nm, B 0·080 nm, C 0·077 nm, N 0·075 nm, O 0·074 nm, and F 0·072 nm. The explanation of this effect lies in the consideration of *effective nuclear charge*. For lithium, the outer electron is screened by two inner electrons, and the effective nuclear charge is $3 - 2 = 1$. For beryllium, the outer electrons are screened by the two inner electrons, giving an effective nuclear charge of 2. Hence, the force exerted on the outer electrons in a beryllium atom is greater, and the atom becomes more compact. This contraction continues across the period.

Energy Changes During Ionic Bond Formation

The processes of ion formation can be separated into the following theoretical stages, for which energy data are available. We take sodium chloride as an example.

Production of the cation *First stage:* Formation of individual gaseous metal atoms from the solid.

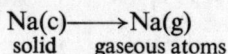

$$Na(c)\longrightarrow Na(g)$$
$$\text{solid} \qquad \text{gaseous atoms}$$

Energy required is the heat of atomization of sodium

$$= 108\cdot8 \times 10^3 \text{ joule}$$
$$= 108\cdot8 \text{ kilojoule (kJ)}$$

Second stage: Conversion of gaseous atoms to ions.

$$Na(g) \rightarrow Na^+ + e^-$$

Ionization energy $= 493\cdot7$ kJ

Production of the anion *First stage:* Formation of individual gaseous non-metal atoms.

$$\tfrac{1}{2}Cl_2 (g) \rightarrow Cl (g)$$

Dissociation energy of chlorine $= 121\cdot3$ kJ

Second stage: Conversion of non-metal atom to an anion.

$$Cl (g) + e^- \rightarrow Cl^-$$

Electron affinity of chlorine $= -359\cdot8$ kJ (energy is liberated)

The total of all these four quantities is the energy needed to produce 1 mole of isolated cations and anions; this equals

$$364\cdot0 \text{ kJ}$$

When the isolated ions are allowed to mix and form a crystal, a considerable amount of energy (known as *crystal* or *lattice energy*) is released. This quantity supplies the energy needed for the above process, while any excess energy is liberated as *heat of formation* of the salt. For sodium chloride, the heat of formation ΔH_f is $-410 \cdot 0$ kJ, so that the size of the lattice energy factor is

$$364 \cdot 0 + 410 \cdot 0 = 774 \text{ kJ/mol}$$

Lattice energy is defined as the energy required to separate completely all the ions in 1 mole of a crystal.

The process is summarized in Figure 12, the cycle of energy changes being called the *Born Haber* cycle.

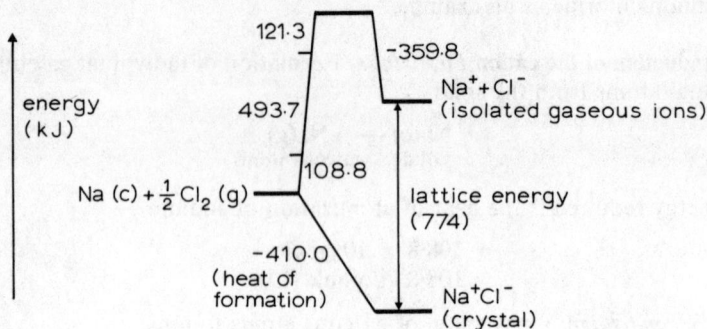

Fig. 12 Born Haber cycle

TABLE 6 *Lattice energies for some common ionic compounds*

Cation	Anion				
	F^-	Cl^-	Br^-	I^-	O^{2-}
Li^+	1·03	0·85	0·84	0·76	2·89
Na^+	0·91	0·77	0·74	0·70	2·52
K^+	0·83	0·70	0·67	0·63	2·23
Mg^{2+}	2·92	2·48	2·39	2·28	3·82
Ca^{2+}	2·63	2·21	2·12	2·03	3·44

Values in MJ/mol.

Some lattice energies are given in Table 6, from which it can be seen that:

(a) lattice energy increases with the size of the charge on the ion,
(b) lattice energy decreases as the size of the ion increases.

This would be expected from the *inverse square law*:

$$\text{Force} = \frac{q_+ \times q_-}{d^2}$$

where q_+ and q_- are the charges carried by the cation and anion respectively, and d is the distance separating the nuclei of the cation and anion.

Factors Governing the Number of Electrons Transferred During Bonding

Knowing (a) the charge carried by the cation and anion, (b) the sizes of the ions, and (c) the manner in which the ions are packed into a crystal, it is possible to calculate a value for the lattice energy of an ionic compound. (The equation used in this connection is the *Born equation*.) Using this technique, and by making reasonable assumptions in regard to ionic size, etc., the energy exchanges involved in the production of 'hypothetical' compounds can be evaluated. Table 7 compares the energetics of producing the non-existent MgCl and $MgCl_3$ with the corresponding values for $MgCl_2$.

TABLE 7 *Heats of formation for MgCl, $MgCl_2$ and $MgCl_3$*

	MgCl ($n = 1$) MJ	$MgCl_2$ ($n = 2$) MJ	$MgCl_3$ ($n = 3$) MJ
Energy to form Mg^{n+} *(isolated, gaseous)*	0·891	2·347	10·075
Energy to form nCl^- *(isolated, gaseous)*	−0·238	−0·477	−0·715
Total	0·653	1·870	9·360
Lattice energy	−0·715	−2·481	−8·535
Heat of formation	−0·062	−0·611	+0·825

From this table, it is clear that the increase in lattice energy fails to meet the very large increase in the energy required to produce an Mg^{3+} ion. Therefore, magnesium (III) chloride is not formed. Comparing MgCl and $MgCl_2$, the extra energy required to produce an Mg^{2+} ion is more than recovered in terms of the increased lattice energy. If the tendency to form a particular compound is measured in terms of the size of the heat of formation, then $MgCl_2$ is the compound most likely to be formed.

To sum up, the number of electrons lost in cation formation (or gained in anion formation) can be deduced from energy considerations instead of the attainment of a 'noble gas structure'. (This approach is useful in connection with the ions formed by the d-block elements.) Since ionization energies and electron affinities must be linked with electronic configurations, this approach means that energy values are used to give an estimate of the degree of stability of the various electronic configurations.

The Importance of Lattice Energy

Lattice energy has already been defined as the energy required to separate completely all the ions present in 1 mole of crystal. Thus, lattice energy is an important factor in connection with the breakdown of a crystal lattice. This may be brought about by:

(a) *Heating.* Salts with high lattice energies require a greater input of thermal energy to break down the crystal lattice. Consequently, such salts have a high melting point. (Examples are: NaF, m.p. 990°C; NaI, m.p. 650°C.)

(b) *Polar solvents.* A water molecule is (i) non-linear, and (ii) polar. The polarity is due to the more electronegative oxygen atom pulling the electrons more closely to itself and away from the hydrogen atoms. This slight separation of charge produces a negative and a positive 'end' in the molecule as shown in Figure 13. When cations or anions are placed in water, the water molecules are attracted to these ions, forming a cluster round them. This process results in a liberation of energy—*hydration energy.* (In general, the energy released by any polar solvent behaving in this way is termed *solvation energy.*) Thus, hydration energy can provide the energy needed to break down the crystal lattice.

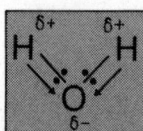

Fig. 13 Polarity of water

General Properties Associated with Ionic Bonding

1 Formation of a large assembly of ions held by electrostatic forces in a crystal lattice.
2 Because of the high lattice energies, ionic compounds have high melting and boiling points.
3 Ionic compounds are usually soluble in polar solvents.
4 Ionic compounds conduct electricity when in solution or when fused, as the ions are then free to move. The conduction of electri-

city is accompanied by decomposition of the compound, since the ions are discharged during this process.

Covalent Bonding

A covalent bond results from the sharing of two electrons between the combining atoms. The atoms approach each other closely, so that the outermost electrons intermingle. A covalent (or *electron pair*) bond therefore involves:

(a) the overlapping and subsequent merging of two atomic orbitals, one from each atom;
(b) the presence of either (i) one unpaired electron in each orbital (the electrons being of opposite spin), or (ii) a pair of electrons in one orbital, the other being empty. The latter constitutes a *coordinate covalent* bond.

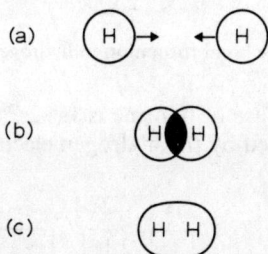

Fig. 14 Stages in the bond formation of a hydrogen molecule

The simplest covalent structure is found in the hydrogen molecule; this is described in Figure 14. The process of formation of the molecule is as follows:

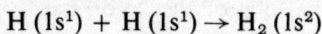

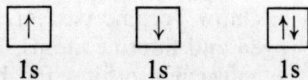

$$H\,(1s^1) + H\,(1s^1) \rightarrow H_2\,(1s^2)$$

The bond formed by this type of overlap is called a σ-bond. In such bonding, rotation of one atom relative to the other, about a line joining the two atoms, does not alter the extent of overlap of the two orbitals. Thus, free rotation is possible about a σ-bond.

The corresponding stages for the formation of a single covalent bond in hydrogen fluoride are shown in Figure 15.

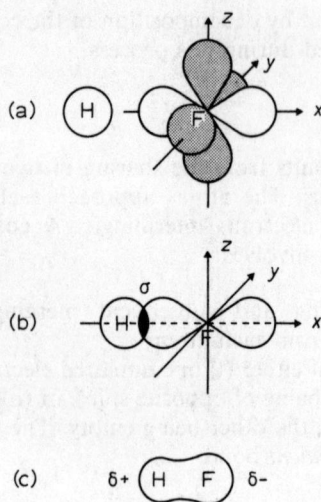

Fig. 15 Single bond formation in hydrogen fluoride

The electronic structure of fluorine is $1s^2\ 2s^2\ 2p^5$, and the vacancy in the $2p_x$-orbital is filled by the hydrogen electron. Again, a σ-bond

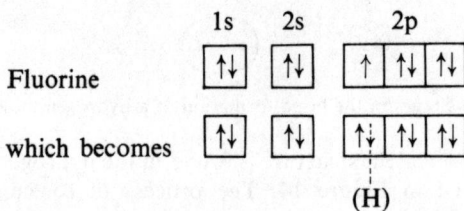

is formed—only one lobe of the 2p-orbital overlaps the hydrogen 1s-orbital. The other lobe may not be used for further overlapping since the 2p-orbital has now become part of a molecular orbital, embracing both hydrogen and fluorine atoms. Unlike the hydrogen molecular orbital, the molecular orbital for hydrogen fluoride is asymmetrical, showing that the two electrons forming the bond are more likely to be found nearer to the fluorine atom. This is due to the fact that fluorine is the more electronegative atom. This brings about a slight separation of charge, leaving the hydrogen end of the molecule slightly positive, and the fluorine end slightly negative. Thus, hydrogen fluoride is a polar molecule.

Similar diagrams may be drawn for the other halogen hydrides.

The bonding process in oxygen is a little more complicated. Oxygen has the configuration $1s^2\,2s^2\,2p^4$, so that there are two p-orbitals which are singly filled. Assuming the $2p_y$-orbital contains two electrons (shaded in Figure 16), a single covalent bond can be formed by the overlap of the p_x-orbitals. The orbitals merge to form a molecular orbital. Now that the oxygen atoms have been

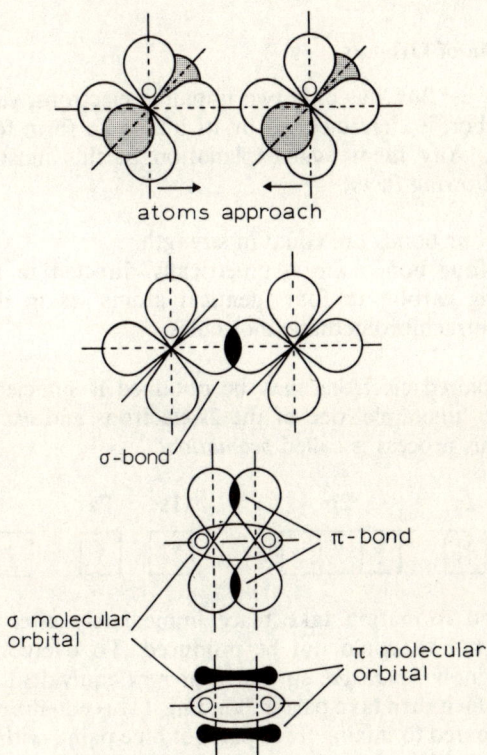

Fig. 16 Bonding in oxygen

drawn close together by the formation of a σ-bond, a second type of overlap—a 'side by side' overlap—of the p_z-orbitals occurs. The upper and lower overlaps, taken together, constitute a new type of single covalent bond, the π-bond, and the π molecular orbital is split into two regions. On rotation of the oxygen atoms, the p_z-orbitals are quickly twisted out of overlap, showing that such rotation would break the bond.

Nitrogen, $1s^2\,2s^2\,2p^3$,

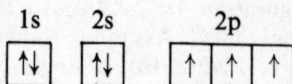

has three singly filled orbitals, and one σ- and two π-bonds are formed, giving rise to a triple bond in the nitrogen molecule.

Hybridization of Orbitals

Carbon, $1s^2\,2s^2\,2p^2$, has only two unpaired electrons, yet the chemistry of carbon is characterized by its ability to form four electron pair bonds. Any theoretical explanation of this must also agree with the following facts:

(a) The four bonds are equal in strength.
(b) The four bonds are symmetrically directed in space when linking carbon to four identical atoms, as in the methane and tetrachloromethane molecules.

Four unpaired electrons may be obtained if sufficient energy is absorbed to 'uncouple' one of the 2s-electrons and excite it to the 2p-level. This process is called *promotion*:

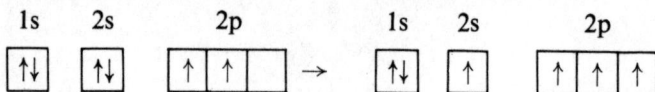

Should bond formation take place immediately after promotion, four equal bonds would not be produced. To overcome this, the orbitals are now *equalized*, so that four new equivalent orbitals are produced which then take part in bonding. (This equalization process can be compared to mixing three parts of blue paint, with one part of yellow. After mixing, four parts of bluish-green paint are formed.) The symbol given to these four new orbitals indicates their origin—

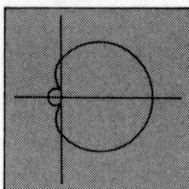

Fig. 17

42

in this case they are labelled sp^3 hybrid orbitals, and the equalization process is termed *hybridization*. The shape of an sp^3 hybrid orbital is shown in Figure 17, and, since there are four of these new orbitals, they are directed symmetrically in space, that is tetrahedrally.

Table 8 summarizes the properties of three sets of hybrid orbitals.

TABLE 8 *Hybrid orbitals*

Parent orbitals	Hybrid orbitals	Number of hybrid orbitals	Spatial arrangement
One s and one p	sp	$1 + 1 = 2$	Linear
One s and two p	sp^2	$1 + 2 = 3$	Plane triangle
One s and three p	sp^3	$1 + 3 = 4$	Tetrahedral

The stages in the bonding of hydrogen and carbon in the formation of methane can be outlined as follows (carbon is $1s^2\,2s^2\,2p^2$):

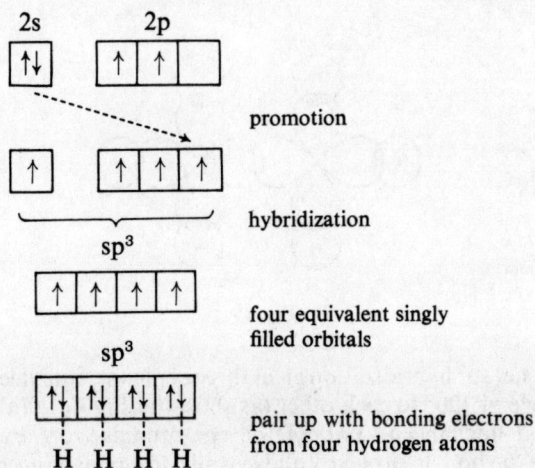

Figure 18 shows the four sp^3-orbitals arranged tetrahedrally. Carbon, when bonded by four single bonds, shows a tetrahedral orientation of these bonds. In diamond, each carbon atom is bonded to four neighbouring carbon atoms by the overlapping of

43

sp³-orbitals, so forming a crystal which is, in effect, a 'giant' molecule.

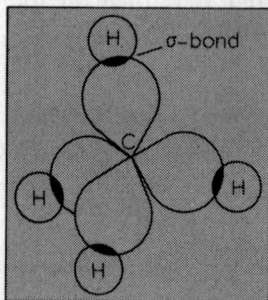

Fig. 18 Tetrahedral orientation of bonds in methane

Bonding in Ethane, Ethene and Ethyne

Tetrahedral (sp^3) hybridization is invoked in describing the bonding in ethane. One sp^3 hybrid orbital (for each carbon) overlaps to form the bond linking the two carbon atoms, and the remaining sp^3 orbitals are engaged in forming the carbon–hydrogen bonds as shown in Figure 19.

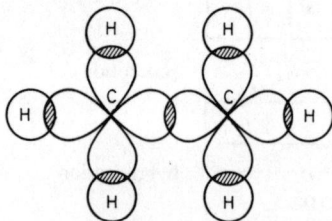

Fig. 19

In ethene, sp^2 hybridization gives three coplanar, equivalent orbitals at an angle of 120° to each other (as shown in Figure 20(a)) with one 2p orbital unchanged. Two of the sp^2 orbitals from each carbon form the carbon–hydrogen bonds, while the remaining sp^2 orbital forms a σ-bond between the two carbon atoms. All these bonds are in the same plane, while the unchanged 2p orbital at right angles to this plane overlaps as shown in Figure 20(b), with the 2p orbital on the other carbon atom to produce the π-bond (the second bond) between the two carbon atoms.

44

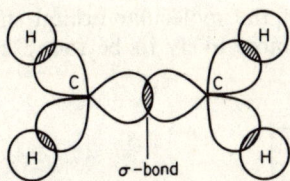

Fig. 20(a)

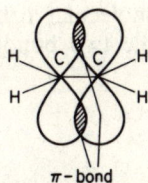

Fig. 20(b)

In ethyne, the 2s and the $2p_x$ orbitals hybridize to form sp hybrids, and these form two collinear sets of bonds (H—C—C—H) as shown in Figure 21(a). Two π-bonds are then formed by the overlap of the unchanged $2p_z$ and $2p_y$ orbitals respectively as shown in Figure 21(b) and (c).

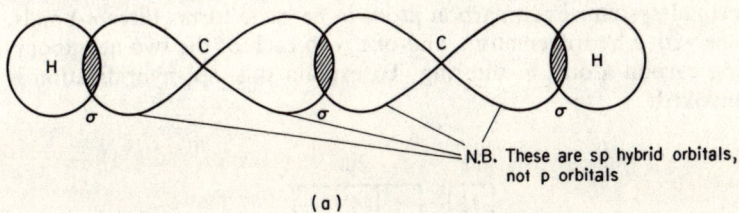

N.B. These are sp hybrid orbitals, not p orbitals

(a)

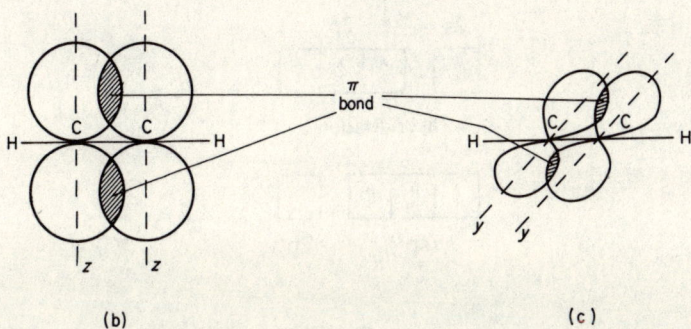

(b)

(c)

Fig. 21

Delocalized Orbitals

The covalent bond uniting two atoms of hydrogen was formed by the overlapping of two atomic orbitals with the subsequent formation of a molecular orbital. The shape of the molecular orbital shows that the two bonding electrons are most likely to be found close

Fig. 22 Localized molecular orbital in hydrogen

to, or in the region between, the two hydrogen nuclei. A molecular orbital which is confined to linking two atoms is known as a *localized* orbital. In the case of benzene, a molecular orbital is formed which envelops six atoms, and this is known as a *delocalized* molecular orbital system. Each carbon atom in benzene forms three σ-bonds, one with a hydrogen atom, and one with each of the two neighbouring carbon atoms in the ring. To explain this, sp^2-hybridization is invoked:

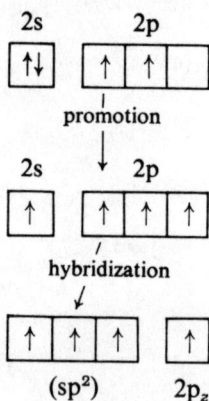

The sp^2-orbitals overlap to form the benzene ring, leaving the p_z-orbital unaffected (Figure 23). The singly filled p_z-orbitals overlap on *each side* to form the large, delocalized molecular orbital system. The six electrons contained in the delocalized orbital are free to move throughout the system, and thus do not belong to any particular carbon atom.

46

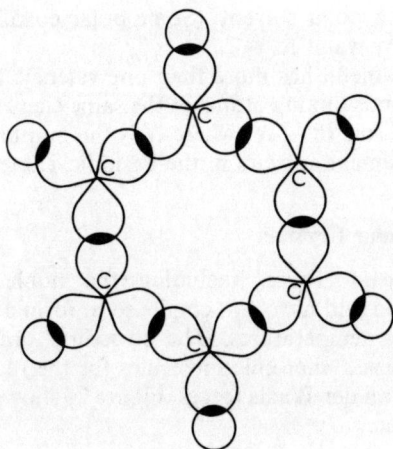

Fig. 23 (sp²) hybrid orbitals in benzene

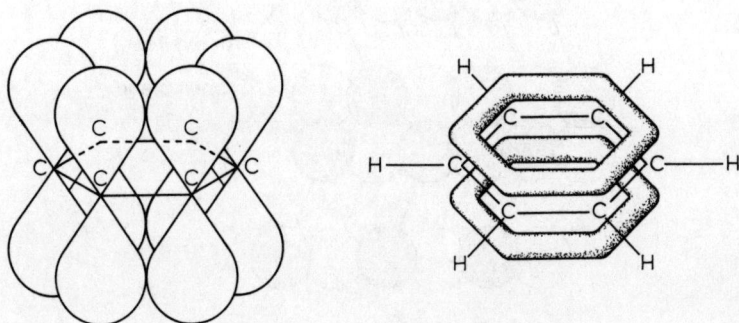

Fig. 24 Formation of a delocalized orbital in benzene

Molecular Crystals

1 The units present in these crystals are discrete molecules.
2 Covalent molecules have no powerful attraction for each other like the ions in an ionic crystal.
3 The individual molecules are held together by weak van der Waals forces (see p. 54). Molecular crystals have, therefore, low melting and boiling points.
4 No charged ions are present. Molecular crystals do not conduct electricity when fused or in solution.
5 The absence of ions minimizes the effect of solvation by a polar solvent. Thus, the solubility of a molecular crystal is not especially

favoured by a polar solvent. (Some polar covalent molecules are hydrolysed by water.)

6 When a non-metal has more than one valency, the lower valency is used in bonds linking atoms of the same element. The valency is given by the rule $(8 - N)$ where N is the number of the group in which the element appears in the Periodic Table.

Typical Molecular Crystals

Gaseous elements These, including the noble gases, fluorine, chlorine, oxygen and nitrogen, condense to form a solid structure at sufficiently low temperatures. The structural units (atoms in the case of noble gases. diatomic molecules for the remainder) are held by very weak van der Waals forces. Figure 25 shows the arrangement in solid chlorine.

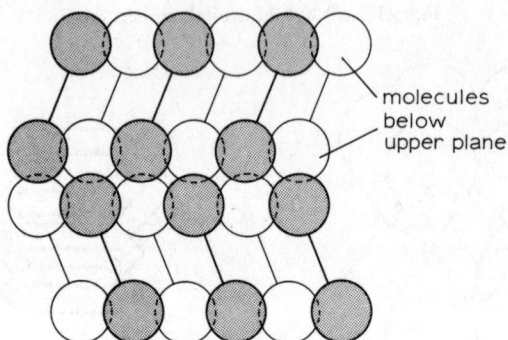

molecules below upper plane

Fig. 25 Chlorine crystal

Sulphur The $(8 - N)$ rule gives the Group VI elements a valency of two. Oxygen forms a doubly bonded diatomic molecule, whereas

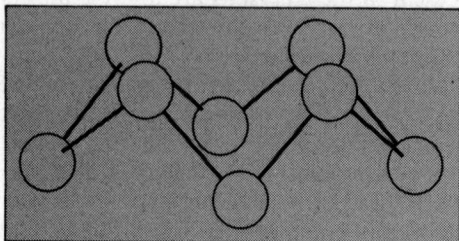

Fig. 26 S_8 unit

48

sulphur forms the cyclic S_8 structure. Covalent bonds link the sulphur atoms in this structure, and the S_8 units are held in place in the crystal by van der Waals forces.

Phosphorus The valency of phosphorus, by the $(8 - N)$ rule, is three, so that each phosphorus atom is bonded to three more in a tetrahedral P_4 molecule. These P_4 units are again attracted by van der Waals forces to form a molecular crystal.

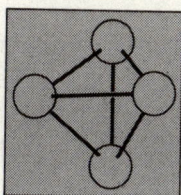

Fig. 27 P_4 molecule

Organic molecules Many form molecular crystals. In addition to van der Waals bonding, hydrogen bonding (see p. 55) occurs and, in some cases, a degree of ionic bonding is also present.

Atomic Crystals

In these crystals, the structural units are linked by covalent bonds and not by van der Waals forces. As a result:

(a) The crystal becomes a 'giant' molecule.
(b) The structural units are held in place by strong bonds; hence the crystals are hard, and have high melting points. They are also insoluble in most solvents.

Examples of atomic crystals follow.

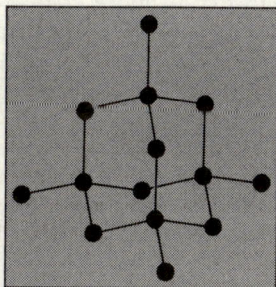

Fig. 28 Diamond structure

49

Diamond Each carbon atom is bonded to four neighbours (the $(8 - N)$ rule) by covalent bonds resulting from sp^3-hybridization, to form a three-dimensional giant molecule.

Graphite Graphite has a layer structure, in which covalent bonds hold the carbon atoms within the layer, while van der Waals forces produce weak inter-layer linkages. This agrees with the ease with which graphite may be sheared along one plane. Alternate layers are identical.

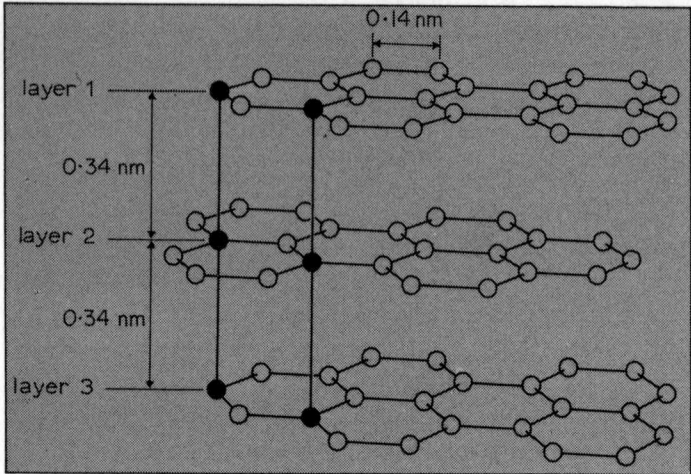

Fig. 29 Graphite

The carbon atoms are bonded (as in benzene) covalently, using sp^2-hybrid orbitals. The p_z-orbitals again form a delocalized system extending above and below each layer. In each delocalized orbital system, the electrons are free to move. This explains the fact that graphite conducts electricity parallel to one plane only.

Metallic Bonding

A molten metal solidifies to produce a crystalline structure, which has a high melting point, high tensile strength, and is capable of conducting heat and electricity. The units are held within a metallic crystal by *metallic bonding*, which may be described as follows:

1 The valence electrons in a metal are not strongly held by the atoms (low ionization energy). The alkali metals have one valence

electron, the Group II metals have two, and so on, while the transition metals can use even more in bond formation (see p. 106).

2 In both benzene and graphite, atomic orbitals may, in certain circumstances, merge to produce a delocalized orbital system, throughout which the electrons are free to move.

3 In a metal, the atomic orbitals holding the valency electrons are assumed to merge in a similar way to produce a giant delocalized system which permeates the whole metal lattice. This delocalized orbital system is the basis of metallic bonding.

4 The valence electrons are free to move in the delocalized system. Under the influence of a potential gradient, an electron flow, that is, a current, is set up. (The conduction of heat is also related to electron flow.)

5 The structural units of a metallic crystal are atoms which have been deprived of their valence electrons (these having entered the delocalized orbital). These units cannot properly be called cations, as no anions are present, and the electrons in the delocalized system are still fairly closely associated with the 'atoms'. The name *atomic core* has been suggested to denote that the unit is a type of cation, in which the charge transfer is less complete than that between a cation and an anion.

6 A metal crystal is, therefore, a close-packed assembly of atomic cores (positively charged) surrounded by an electron flux. (A crude two-dimensional analogy is the close-packed arrangement adopted when a number of glass spheres are placed on a watch glass which is gently shaken. The spheres represent the atomic cores, while the attractive force of the electron flux may be compared to the force of gravity on the spheres due to the concave surface of the watch glass.)

7 The strength of a metal bonding system increases with:

(a) an increase in the number of electrons in the delocalized system,

(b) a decrease in the size of the atomic core which forms the structural unit.

The alkali metals, which form large cations, have only one electron which is available to enter the delocalized system. The strength of the metallic bond is low, and these metals have low melting points and are soft, malleable and ductile.

Transition metals form small cations, and can release quite a number of electrons into the metallic bond, so that these metals are hard, with a high tensile strength, and have a high melting point, corresponding to a strong metallic bonding system.

51

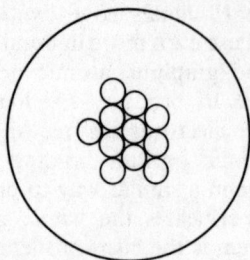

Fig. 30 Close-packed spheres in a watch glass

Metallic Crystals

The structural units—the positively charged atomic cores—may be regarded as uniform spheres which pack together as closely as possible in the crystal. Close packing of spheres is shown in Figure 31(a) and (b). The first layer is formed by placing the spheres as shown in (a), and the second layer is added by placing the spheres in the hollows formed by three spheres in the first layer. A third layer may be added in one of two ways:

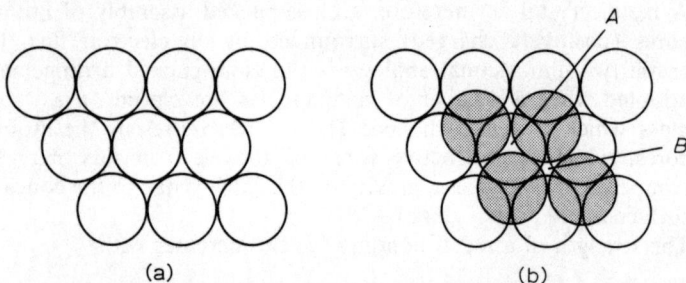

Fig. 31 Close packing of uniform spheres

(i) Spheres are placed in the hollows marked *A*, making the first and third layers identical. The resulting structure is known as *hexagonal close packing*. Magnesium and zinc crystallize with this type of structure.

(ii) Spheres are placed in the hollows marked *B*, making the third layer different from both preceding layers. The resultant structure is called *face-centred cubic packing*. Copper, iron, silver and calcium have face-centred cubic lattices.

A third type of packing results from a first layer which is not as close-packed as that shown in Figure 31. The addition of successive

52

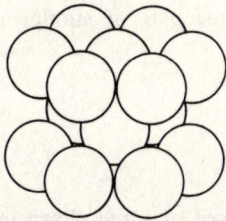

Fig. 32 Hexagonal close packing

Fig. 33 Face-centred cubic

layers of spheres resting in the hollows formed by *four* spheres in the preceding layer, produces a *body-centred cubic* structure. The alkali metals have body-centred cubic crystals.

Fig. 34 Body-centred cubic

Intermediate Types of Bond

In a covalent bond between like atoms, the electrons are shared equally, as in the hydrogen molecule (p. 39). When a covalent bond links two different atoms, the electron sharing is unequal, as in the case of hydrogen fluoride (p. 40). The electrons are attracted towards the more electronegative atom (fluorine) and away from the hydrogen, so that a dipole is produced. The polarity of a molecule such as hydrogen fluoride is indicated by writing

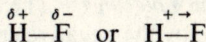

$$\overset{\delta +}{H} - \overset{\delta -}{F} \quad \text{or} \quad \overset{+ \rightarrow}{H - F}$$

where the arrowhead shows the negative end of the molecule. Consequently, covalent bonds have a certain degree of ionic character, which can be expressed in terms of a dipole moment μ, which is found by taking the product of the size of the charges and the distance between them. For hydrogen fluoride, μ is 6.3×10^{-30} C m whereas hydrogen chloride has a dipole moment of 3.45×10^{-30} C m

Polar molecules show unusual properties:

(a) They are associated, since they attract one another.
(b) They have higher boiling points than would be expected from

53

a comparison with non-polar molecules of similar relative molecular mass.

(c) They have a high dielectric constant.

(d) They are good ionizing solvents.

Fajans' Rules

Just as covalent bonds show some degree of ionic character, it is found that the charge separation between a cation and an anion may not be complete. Fajans pointed out that a small, highly charged cation is strongly polarizing (i.e., has the power to distort, or polarize, the electron screen of a neighbouring atom or anion), while a bulky anion does not hold its outer electrons very firmly, so that these anions are easily polarized. In this way, a cation may retrieve a share in the electron (or electrons) lost on ionization, and the ionic bond may be regarded as having a certain covalent content. Fajans outlined three rules which summarized the situations in which an ionic bond shows the greatest covalent character.

1 When the cation is small.
2 When the anion is large.
3 When the charge on both anion and cation is high.

The properties of lithium (p. 97) and aluminium (p. 132) illustrate these rules.

Van Der Waals Forces

These weak forces which link covalent molecules (or noble gas atoms) may originate in three ways:

1 *Attraction between polar molecules.* The attraction of the positive end of a polar molecule by the negative end of another produces a van der Waals force.

2 *Attraction between polar and non-polar molecules.* Just as a magnet can attract a piece of soft iron by inducing in it a pole of opposite sign, a polar molecule can produce a similar effect on a non-polar molecule. To respond to the polarizing effect of the polar molecule, the non-polar molecule must be easily polarizable.

3 *Dispersion forces.* The electrons surrounding the nucleus of an atom are in constant motion, and it is possible that, at any given instant, the nucleus will be less well screened in one direction, and over-screened in another. This allows for the formation of *instantaneous* or *temporary* dipoles. As one dipole disappears,

another may form, thus providing a continued attractive force between the atoms or molecules.

Attractions resulting from van der Waals forces can be regarded as constituting a very weak bond between neighbouring atoms, ions or molecules. The energy required to break this type of bond is of the order of 5 kJ mol^{-1} (compared with 436 kJ mol^{-1} to break the covalent bond in a hydrogen molecule, or 775 kJ mol^{-1} which is the lattice energy of sodium chloride). Because of van der Waals forces, gases, such as hydrogen, oxygen and argon, liquefy and solidify, while molecules, such as naphthalene, crystallize. Since the van der Waals forces are easily overcome, the condensed gases are readily vaporized, and molecular crystals are soft and have low melting points. Van der Waals forces hold the discrete molecules together in a liquid.

Hydrogen Bonding

When hydrogen is covalently bonded to a more electronegative element, the bond is polar, and a dipole moment is set up:

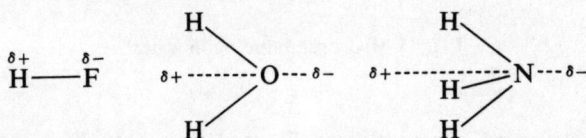

The slightly positive hydrogen atom in these molecules is capable of attracting the negative end of a neighbouring molecule, and so on, so that the molecules become associated. The strength of this attraction is ten times that of a van der Waals bond, and about one-tenth of that of a normal covalent bond. This type of attractive force is called a *hydrogen bond*. Hydrogen bonding is responsible for such features as the following:

Higher boiling and melting points   Examples are:

   CH_4 (no hydrogen bonding, relative molecular mass (M_r) 16),
      boiling point $-161°C$,
   H_2S (a little hydrogen bonding, M_r 34), boiling point $-60°C$,
   H_2O (extensive hydrogen bonding, M_r 18), boiling point $100°C$.

In the absence of hydrogen bonding, the boiling point of water would be expected to be below that of hydrogen sulphide, and almost as low as that of methane. Similarly, ammonia boils at −33°C and phosphine at −87°C. A comparison of melting points also indicates a similar trend.

Structure in the solid state Hydrogen bonding (shown by the dotted lines in Figure 35) links the molecules in a regular pattern, and consequently confers a regular structure on these substances when in the solid state. Water is the best example of this effect.

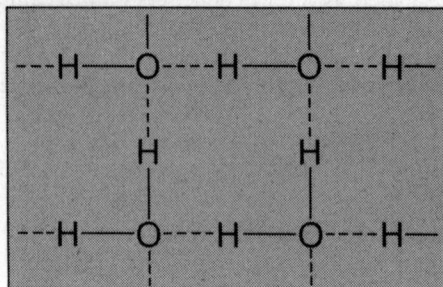

Fig. 35 Hydrogen bonding in water

The structure, shown flat in Figure 35, is actually tetrahedral, and is very open. When ice melts, this open structure is broken down to form a more compact assembly, so that liquid water is less dense than ice. Liquid water at 0°C shows a degree of hydrogen bonding, but this is broken down by thermal agitation as the water is heated slightly. This explains why water continues to contract on heating from 0°C to 4°C, after which natural thermal expansion outweighs the contraction due to the rupture of the remaining hydrogen bonds.

Hydrogen bonding between hydrogen and fluorine accounts for the formation of the ion HF_2^- in the solid salts (e.g., KHF_2). The ion is linear with equal hydrogen–fluorine bond lengths.

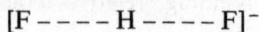

$$[F----H----F]^-$$

High latent heats of fusion and vaporization are associated with hydrogen bonding, and this effect is also responsible for the higher boiling points of organic acids and alcohols over aldehydes and ethers of similar relative molecular mass. Some values are given in Table 9.

56

TABLE 9 *Relationship between boiling points and hydrogen bonding*

Ethanoic acid	b.p.	391K
Propanoic acid	b.p.	414K
Ethanol	b.p.	352K
Propanol	b.p.	371K
Ethanal	b.p.	294K
Propanal	b.p.	322K
Methoxymethane	b.p.	248K
Ethoxyethane	b.p.	308K

Electronegativity

The dipole moment of hydrogen fluoride is 6.3×10^{-30} C m, while that of hydrogen chloride is 3.45×10^{-30} C m. This indicates that fluorine has a greater electron-attracting power than chlorine. In order to compare the ability of various elements to attract electrons, a scale of electronegativity values was drawn up by Pauling. The *electronegativity x* of an element is a measure of the electron-attracting power of an element *when combined in a molecule*. Table 10 gives the electronegativity values for a number of elements. No electronegativity values have been assigned to the noble gases.

TABLE 10 *Electronegativities*

Li	Be	B	C	N	O	F	H
1	1·5	2	2·5	3	3·5	4	2·1
Na	Mg	Al	Si	P	S	Cl	
1	1·2	1·5	1·7	2·1	2·4	2·8	
			Ge	As	Se	Br	
			2·0	2·2	2·5	2·7	
				Sb	Te	I	
				1·8	2·0	2·2	

Electronegativity values show the following general trends:

(a) they increase from left to right across a period (this corresponds to a decrease in the size of the atom);
(b) they decrease with increasing atomic number within a group.

Fluorine is the most electronegative element, while caesium (with a value of 0·8) is the least.

The usefulness of the electronegativity concept will become apparent in later sections, although an immediate application is in the estimation of the polarity of a bond. Bonds formed between elements of widely different electronegativities are mainly ionic—when $x_A - x_B$ is about 1·8 or more, the bond is thought to be at least 50 per cent ionic. For example,

$$x_F - x_{Al} = 2·5 \quad \text{and} \quad x_{Cl} - x_{Al} = 1·3$$

indicating that aluminium fluoride is ionic, while aluminium chloride is more covalent.

Electronegativity values are also useful in the interpretation of the hydrolysis products of covalent halides, in understanding the shapes of molecules in terms of the electron pair repulsion theory (see p. 139), and in the assignment of oxidation numbers (see p. 12).

5 Chemical Reactions

A chemical reaction produces new substances (*products*) from suitable starting materials (*reactants*).

The fact that a chemical reaction has taken place is shown by one or more of the following factors:

(a) A colour change is observed.
(b) A precipitate is produced, or a gas is evolved.
(c) A change in pH can be detected.
(d) Heat is given out or absorbed.

Every chemical reaction is accompanied by:

1 a heat (or energy) change,
2 a redistribution of matter,

and the combined effects of these two factors is expressed by the quantity *free energy* (symbol *G*).

We will consider each factor in turn.

Heat Changes

Both the reactants and products can be assumed to possess a store of energy (made up of kinetic energy of motion of the atoms and molecules, bond energy, lattice energy, etc.). The total energy contained within each substance is called the *heat content* or *enthalpy* of that substance. When a reaction takes place, the enthalpies of the products will differ from those of the reactants. If the heat content of the products is lower than the heat content of the reactants, as shown in Figure 36, heat is liberated when the reaction takes place. The amount of heat liberated is denoted by ΔH, where

$$\Delta H = H_{\text{products}} - H_{\text{reactants}}$$

Note that the symbol Δ signifies a change in value, and is always calculated by subtracting the initial quantity from the final quantity.

One might expect that the reactions most likely to take place would be those accompanied by an evolution of energy (i.e., transitions from a higher to a lower energy level), just as it is natural to expect water to flow downhill, losing potential energy in the process. In fact, very many chemical reactions are exothermic, and follow this expected pattern.

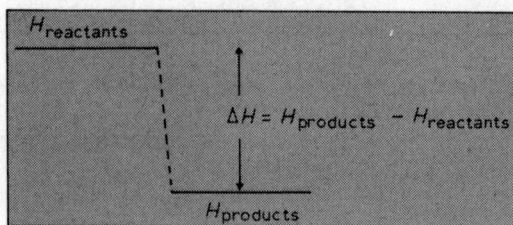

Fig. 36 Heat changes during an exothermic reaction

Redistribution of Matter

Some reactions do however take place against the 'energy gradient' shown in Figure 36. The reason for this can be found by considering the second factor—the redistribution of matter.

The kinetic theory proposes that particles of matter are in constant motion. In a solid, the motion is restricted to a vibration of structural units about a mean position, but with liquids, and especially in gases, the particles are capable of random movement. Any system in random motion tends to become more 'mixed up' or disorderly, as time passes. Very many examples of this can be quoted—a pack of cards originally sorted into suits becomes disordered on shuffling, separate layers of salt and sand intermingle when shaken in a tube, etc.

A measure of the disorder in a system is known as its *entropy* (symbol S), and changes in entropy are designated by ΔS. Changes which are accompanied by a large increase in entropy are:

(a) Change of state (i) solid to liquid, (ii) liquid to gas, (iii) solid to gas.
(b) Reactions in which a gas is produced by the decomposition of a liquid or a solid.

Note that in these changes, the degree of disorder in the final state is greater than that of the initial state, so that ΔS, which is given by

$$\Delta S = S_{\text{final}} - S_{\text{initial}}$$

is positive.

60

S and ΔS are measured in units of joules per degree (absolute), while both ΔH and ΔG values are given in joules. Hence, in the equation connecting ΔS, ΔG and ΔH, the ΔS values must occur multiplied by the absolute temperature T at which the reaction takes place, in order that all the quantities are expressed in joules. The equation is

$$\Delta G = \Delta H - T . \Delta S$$

A reaction is expected to take place if:

(a) heat is evolved (that is, ΔH is negative), and
(b) there is an increase in entropy (that is, ΔS is positive).

Applying these conclusions to the equation above, we see that *a reaction is theoretically possible when ΔG is negative*. For endothermic reactions (heat absorbed, ΔH positive), the overall value of ΔG may be negative, if a large increase in entropy accompanies the reaction.

Some examples of the influence of these factors can be seen in the following reactions.

1 For the reaction

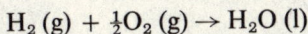

$$H_2 \text{ (g)} + \tfrac{1}{2}O_2 \text{ (g)} \rightarrow H_2O \text{ (l)}$$

ΔH is -285 kJ.

On forming 1 mole of liquid water from $1\frac{1}{2}$ moles of gaseous reactants, the system has become more ordered, and entropy has been lost. Since we know that this reaction takes place the effect of a decrease in entropy must be a minor factor, as ΔG must remain negative. Actual values are:

$$\begin{aligned} \Delta H &= -285 \text{ kJ} \\ T . \Delta S &= + \ 48 \text{ kJ} \\ \hline \Delta G &= -237 \text{ kJ} \end{aligned}$$

2 In the reaction

$$N_2O_4 \text{ (g)} \rightarrow 2NO_2 \text{ (g)}$$

heat is absorbed as nitrogen dioxide is formed. However, 2 moles of gas are produced by the decomposition of 1 mole of gas, and

61

there is an increase in entropy sufficient to give ΔG a negative value.

3 In the decomposition of ammonium dichromate (VI)

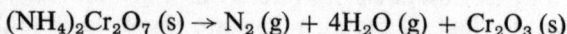

$$(NH_4)_2Cr_2O_7 \text{ (s)} \rightarrow N_2 \text{ (g)} + 4H_2O \text{ (g)} + Cr_2O_3 \text{ (s)}$$

both factors operate to make ΔG negative, since heat is evolved, and there is a large increase in entropy corresponding to the formation of 5 moles of gas from 1 mole of solid reactant.

In general, chemical reactions do not occur spontaneously—for example magnesium ribbon requires heating before it will ignite in air, and a mixture of hydrogen and chlorine does not react until exposed to light. In order to initiate a reaction, a certain amount of energy, known as *activation energy* (see Part 3), must be supplied to the reactants. In some cases the activation energy is large, while in others it is small. The reactions shown in Figure 37 involve the same

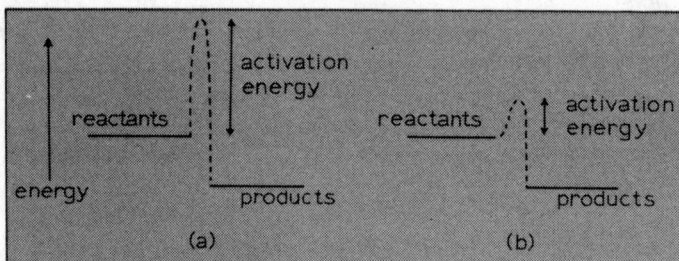

Fig. 37 Activation energy. (a) Large (b) Small

decrease in energy between initial and final states, but, because of the high activation energy required, reaction (a) is much slower than reaction (b).

To sum up:

1 The possibility of a reaction taking place depends on ΔG values.
2 The speed at which a reaction takes place depends on the value of the activation energy.

An obvious consequence of this is that a reaction which may be thought possible (ΔG being negative), cannot always be effected, since the speed of the reaction may be infinitesimally small.

62

Types of Reaction

Neutralization This is the reaction between an acid and a base, to produce a salt and water. An example is:

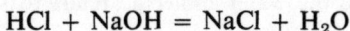

$$HCl + NaOH = NaCl + H_2O$$

or, in general;

$$H^+ + OH^- \rightarrow H_2O$$

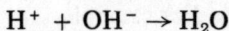

An acid salt results from the partial replacement of the acidic hydrogen atoms (e.g., $NaHSO_4$), while a basic salt can be regarded as the partial neutralization of a base by an acid (e.g., $Pb(OH)NO_3$). Hydroxides of metals, basic oxides and carbonates are all capable of neutralizing an acid:

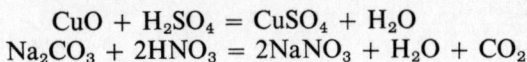

$$CuO + H_2SO_4 = CuSO_4 + H_2O$$
$$Na_2CO_3 + 2HNO_3 = 2NaNO_3 + H_2O + CO_2$$

Neutralization reactions do not involve any change in the oxidation number of any of the elements taking part in the reaction.

Exchange of ions (double decomposition) In these reactions, a product is displaced from the reaction medium as a precipitate. For example,

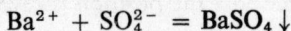

$$Ba^{2+} + SO_4^{2-} = BaSO_4 \downarrow$$

This type of reaction may be extended to include displacements which are not strictly an exchange of ions, but where a product separates as a distinct phase; for example,

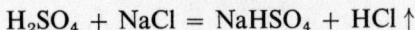

$$H_2SO_4 + NaCl = NaHSO_4 + HCl \uparrow$$

Thermal decomposition The action of heat on most compounds brings about a degree of decomposition. The specific effects of heat on salts such as nitrates and carbonates will be described in later sections.

Some thermal decompositions are reversible:

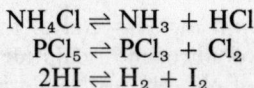

$$NH_4Cl \rightleftharpoons NH_3 + HCl$$
$$PCl_5 \rightleftharpoons PCl_3 + Cl_2$$
$$2HI \rightleftharpoons H_2 + I_2$$

Hydration This is the formation of a product by the addition of (and bonding of) water to another reactant. Examples are:

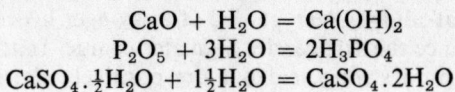

$$CaO + H_2O = Ca(OH)_2$$
$$P_2O_5 + 3H_2O = 2H_3PO_4$$
$$CaSO_4 . \tfrac{1}{2}H_2O + 1\tfrac{1}{2}H_2O = CaSO_4 . 2H_2O$$

63

Hydrolysis These are reactions in which a substance is decomposed by the action of water. The products of hydrolysis of many covalent compounds can be understood in terms of electronegativities, since, in these reactions, the water molecule tends to split $\overset{\delta+}{H}$—$\overset{\delta-}{OH}$. The $\overset{\delta-}{—OH}$ end of the water molecule reacts with the positive end of the polar bond under attack. For example,

$$
\begin{array}{c}
\text{Cl} \quad \overset{\delta-}{\text{Cl}} \qquad \overset{\delta+}{H} \\
\diagdown \diagup \qquad \diagup \\
\text{P}\,\delta+ \quad + \quad \overset{}{\text{OH}} \qquad \longrightarrow \qquad
\begin{array}{c}
\text{Cl} \\
\diagdown \\
\text{P—OH} + \text{HCl} \\
\mid \\
\text{Cl}
\end{array} \\
\mid \qquad\qquad \underset{\delta-}{} \\
\text{Cl}
\end{array}
$$

$$\downarrow \begin{array}{l} 2H_2O \\ \text{(further hydrolysis)} \end{array}$$

$$H_3PO_3 + 3HCl$$

The electronegativity of chlorine is greater than that of phosphorus, so that the phosphorus atom is slightly positive, and reacts with the $\overset{\delta-}{—OH}$ end of the water molecule.

With nitrogen trichloride, the position is reversed:

$$
\begin{array}{c}
\text{Cl} \quad \overset{\delta+}{\text{Cl}} \qquad \overset{\delta-}{\text{OH}} \\
\diagdown \diagup \qquad \diagup \\
\text{N}\,\delta- \quad \overset{\delta+}{H} \quad \longrightarrow \quad
\begin{array}{c}
\text{Cl} \\
\diagdown \\
\text{N—H} + \text{HOCl} \\
\mid \\
\text{Cl}
\end{array} \\
\mid \\
\text{Cl}
\end{array}
$$

$$\downarrow \begin{array}{l} 2H_2O \\ \text{(further hydrolysis)} \end{array}$$

$$NH_3 + 3HOCl$$

Here, the nitrogen atom, being more electronegative, is slightly negative, and reacts with the $\overset{\delta+}{H}$— end of the water molecule. The hydrolysis products of nitrogen trichloride are thus different in character from those of phosphorus trichloride.

Hydrolysis is a reaction of particular importance in organic chemistry, where, in addition to water, alkalies such as sodium hydroxide, or acids such as sulphuric, are used to bring about this reaction. An alkali may be regarded as a stronger hydrolysing agent than water, since the OH carries a greater charge. In the same way, the hydrogen in a strong acid is more positively charged than the hydrogen in water, which makes the acid a more effective hydrolysing agent than water.

Oxidation and reduction (redox) reactions The definition of oxidation has undergone progressive modification. Originally, it meant the combination of an element with oxygen, as in the reaction

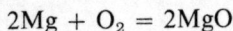

$$2Mg + O_2 = 2MgO$$

Here the oxidation number of the magnesium increases from zero (in the metallic state) to $+2$ (in the oxide), while the oxidation number of oxygen decreases from zero to -2.

An alternative approach is to consider the reaction in two parts:

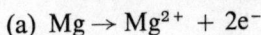

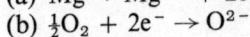

(a) $Mg \rightarrow Mg^{2+} + 2e^-$
(b) $\frac{1}{2}O_2 + 2e^- \rightarrow O^{2-}$

In these half reactions, oxidation of magnesium corresponds to the loss of electrons, while the oxygen which is reduced accepts electrons. The modern definitions of oxidation and reduction are framed in terms of change of oxidation state, or in terms of electron transfer:

Oxidation is a process which causes an increase in the oxidation number of a substance, or a process in which a species loses electrons.
Reduction is a process in which the oxidation number of a species is reduced, or a process in which a substance accepts electrons.

Oxidation numbers are useful in deciding whether a substance has been oxidized or reduced during a reaction. For instance,

$$Mg + 2HCl = MgCl_2 + H_2$$

The oxidation number of magnesium, initially zero, is changed to $+2$ on the dissolution of the metal in acid. Magnesium has therefore been oxidized during this reaction. The corresponding reduction takes place with hydrogen, whose oxidation number changes from $+1$ in the acid, to zero in hydrogen gas. Note that in this sense, all acids which dissolve a metal to produce metal ions, are oxidizing acids.

It is convenient to write the oxidation numbers for an element (or an ion) below the symbols in an equation:

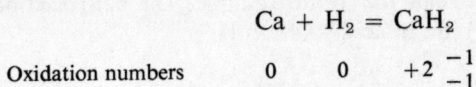

$$Ca + H_2 = CaH_2$$

Oxidation numbers   0   0 $+2$ $\begin{matrix} -1 \\ -1 \end{matrix}$

Thus, calcium is oxidized and hydrogen is reduced in the above reaction.

As a second example,

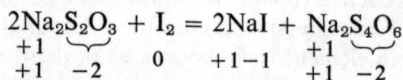

$$2Na_2S_2O_3 + I_2 = 2NaI + Na_2S_4O_6$$

In this reaction, iodine has been reduced and the thiosulphate ion $S_2O_3^{2-}$ has been oxidized to a tetrathionate ion $S_4O_6^{2-}$.

Another type of redox reaction is the following:

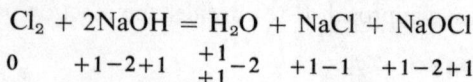

$$Cl_2 + 2NaOH = H_2O + NaCl + NaOCl$$

Here chlorine is both oxidized to a $+1$ oxidation state in the chlorate (I) ion OCl^-, and reduced to a -1 oxidation state in the chloride ion Cl^-. This type of reaction in which the same species undergoes simultaneous oxidation and reduction is known as *disproportionation*.

Another example of disproportionation is the decomposition of copper (I) sulphate in the presence of water:

$$Cu_2SO_4 = CuSO_4 + Cu\downarrow$$

Balancing Redox Equations

The reduction and oxidation reaction, or redox reaction, can be considered in two parts, called half reactions, in terms of electron transfer. For example,

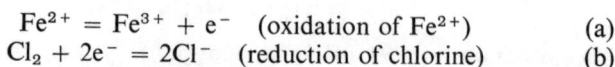

$$Fe^{2+} = Fe^{3+} + e^- \quad \text{(oxidation of } Fe^{2+}) \qquad \text{(a)}$$
$$Cl_2 + 2e^- = 2Cl^- \quad \text{(reduction of chlorine)} \qquad \text{(b)}$$

To obtain the balanced reaction between iron (II) ion and chlorine, equation (a) must be multiplied by 2, so as to make the supply of electrons equal to the demand in equation (b). Adding $2 \times$ (a) to (b),

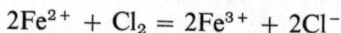

$$2Fe^{2+} + Cl_2 = 2Fe^{3+} + 2Cl^-$$

Some common oxidizing agents are oxy-anions of the transition metals; potassium manganate (VII), for example. The half reaction for the manganate (VII) ion in acid solution is

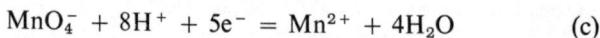

$$MnO_4^- + 8H^+ + 5e^- = Mn^{2+} + 4H_2O \qquad \text{(c)}$$

66

(Note: the number of hydrogen ions required on the left of the equation is twice the number of oxygen atoms in the anion. The transition metal ion on the right of the equation appears in its normal oxidation state—although the 'normal' oxidation state may vary according to the pH of the solution. The charges on each side of the equation are balanced by the addition of electrons, five of which are needed in this case.)

The corresponding equation for dichromate (VI) ion in acid solution is

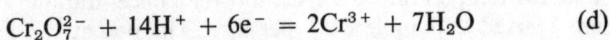

$$Cr_2O_7^{2-} + 14H^+ + 6e^- = 2Cr^{3+} + 7H_2O \qquad (d)$$

Ethanedioc acid and ethanedioates are reducing agents:

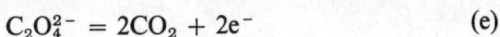

$$C_2O_4^{2-} = 2CO_2 + 2e^- \qquad (e)$$

The balanced equation between an ethanedioate and manganate (VII) in acid solution is found by multiplying equation (c) by 2, and equation (e) by 5, to make the supply of electrons equal to the demand. Adding $2 \times$ (c) to $5 \times$ (e) gives

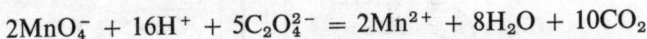

$$2MnO_4^- + 16H^+ + 5C_2O_4^{2-} = 2Mn^{2+} + 8H_2O + 10CO_2$$

In a similar way, $6 \times$ (a) + (d) gives the balanced equation for the reaction between dichromate (VI) ion and iron (II) in acid solution:

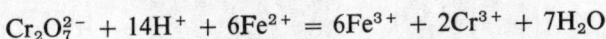

$$Cr_2O_7^{2-} + 14H^+ + 6Fe^{2+} = 6Fe^{3+} + 2Cr^{3+} + 7H_2O$$

Measurement of Oxidizing Power

Oxidation is the loss of electrons by a species. One of the simplest examples of this is the formation of cations by a metal, and the ability of various metals to ionize can be compared by measuring their standard electrode potentials.

When a metal bar is placed in a solution of its ions, a potential difference is established between the bar and the solution. This is a result of the metal atoms losing electrons, and forming cations. Thus,

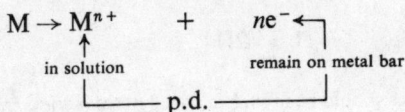

$$M \rightarrow M^{n+} + ne^-$$

in solution   remain on metal bar

p.d.

67

The size of the potential difference developed depends on:

(a) the metal used,
(b) the concentration of the metal ions already in solution,
(c) the temperature.

To enable a comparison of the tendency of various metals to lose electrons to be made, only factor (a) can be variable, and factors (b) and (c) have to be fixed at chosen standard values. The accepted standard temperature is 25°C, and the concentration* of the metal ions is fixed as 1 mole M^{n+} per litre. The potential set up between the metal electrode and a reference electrode (which is a standard hydrogen electrode) is the *standard electrode potential* (designated $E^{\ominus}$) for the metal.

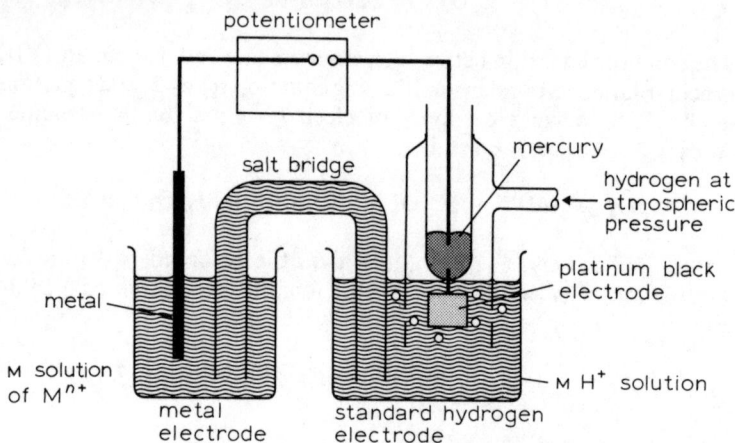

Fig. 38 Measurement of standard electrode potentials

The apparatus used is shown in Figure 38. The standard hydrogen electrode consists of a platinum electrode coated with platinum black, which is immersed in a solution of acid (concentration 1 mole H^+ per litre) and around which hydrogen gas is bubbling under 1 atmosphere pressure, so that the electrode is in contact with both acid and gas.

Hydrogen ionizes according to the equation

$$H_2 \rightleftharpoons 2H^+ + 2e^-$$

* Strictly, activities should be used. See Part 3 of this series, *Physical Chemistry*.

the hydrogen molecules being adsorbed by the platinum black. The two electrodes are coupled by a potentiometer, and the return connection is made by means of a salt bridge (a tube filled with concentrated potassium chloride solution); this arrangement allows the potential difference between the two electrodes to be measured.

The potential of the hydrogen electrode is arbitrarily chosen as zero. Two cases may arise:

(a) The metal M loses electrons (so forming cations) more readily than hydrogen. Thus, electrons accumulate on the metal bar to a greater extent than on the hydrogen electrode. The metal consequently acquires a negative potential with respect to the hydrogen electrode.

(b) The metal M forms ions less readily than hydrogen. In this case, the metal electrode appears positive with respect to the hydrogen electrode.

Standard electrode potentials have been measured for most elements that form ions in solution. For the very reactive metals (e.g., sodium) which react with water, special techniques, such as using a sequence of solutions of various concentrations of the metal in mercury, are necessary.

An arrangement of the elements in the order of their standard electrode potentials is known as the *electrochemical series*; a shortened version is given in Table 11.

A negative $E^{\ominus}$ value means that the element loses electrons more readily than hydrogen—it is a better reducing agent than hydrogen.

TABLE 11 *Standard electrode potentials*

Metal electrode	$E^{\ominus}$	
$K^+ + e^- = K$	$-2 \cdot 92$ V	
$Ca^{2+} + 2e^- = Ca$	$-2 \cdot 87$ V	
$Na^+ + e^- = Na$	$-2 \cdot 71$ V	Very reactive metals
$Mg^{2+} + 2e^- = Mg$	$-2 \cdot 37$ V	
$Al^{3+} + 3e^- = Al$	$-1 \cdot 67$ V	
$Zn^{2+} + 2e^- = Zn$	$-0 \cdot 76$ V	Moderately reactive
$Fe^{2+} + 2e^- = Fe$	$-0 \cdot 44$ V	metals
$Pb^{2+} + 2e^- = Pb$	$-0 \cdot 13$ V	
$H^+ + e^- = \frac{1}{2}H_2$	$0 \cdot 00$ V	Reference standard
$Cu^{2+} + 2e^- = Cu$	$+0 \cdot 34$ V	Noble metal
$\frac{1}{2}I_2 + e^- = I^-$	$+0 \cdot 54$ V	
$Ag^+ + e^- = Ag$	$+0 \cdot 80$ V	Noble metal
$\frac{1}{2}Br_2 + e^- = Br^-$	$+1 \cdot 07$ V	

A positive $E^\ominus$ value means that the element is a poorer reducing agent than hydrogen.

Note that the electrochemical series is a relative sequence. Whether the sign of the $E^\ominus$ value for a metal is positive or negative (indicating that it forms ions less readily, or more readily, than hydrogen), all the metals are *electropositive*, that is they form positive ions.

The equations corresponding to the electrode reaction are written so that the electrons, and the oxidized state of the element, appear on the left of the equation. This is an important conventional point.

Standard electrode potentials are very useful in relation to the chemistry of the metals (see pp. 74–136).

Oxidation–Reduction (Redox) Potentials

The apparatus shown in Figure 38 can be modified to measure the tendency for electron transfer to take place between two ionic species in different oxidation states (e.g. Fe^{3+} and Fe^{2+}), instead of between a metal and its ions.

The metal bar dipping into a solution of its ions is replaced by a platinum electrode which dips into a solution containing the two ionic species, as shown in Figure 39. This electrode is called a

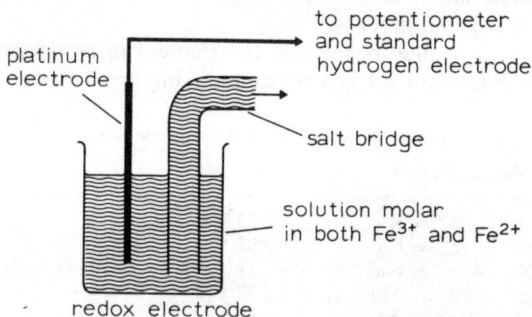

Fig. 39 Measurement of redox potentials

redox electrode, and it is again coupled with a standard hydrogen electrode. The potential difference developed between the two electrodes is known as the *oxidation–reduction potential* (or, more commonly, the *redox potential*) of the couple.

Again, the potential difference for a given redox couple depends on several factors:

(a) temperature,
(b) concentration of the oxidized and reduced forms of the couple,
(c) the pH of the solution.

A standard state is again defined, as follows:

(a) the temperature is kept constant at 25°C,
(b) the concentration of both oxidized and reduced forms is 1 mole per litre,
(c) the pH is zero.

Under these conditions, the measured potential is the $E^{\ominus}$ value, that is the standard redox potential for the couple.

The $E^{\ominus}$ value for the Fe^{3+}/Fe^{2+} system is $+0.76$ V. This is written in equation form as follows:

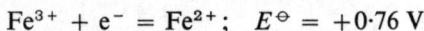

$$Fe^{3+} + e^- = Fe^{2+}; \quad E^{\ominus} = +0.76 \text{ V}$$

The method of writing the equation is in accord with the convention mentioned on p. 70—the electrons and the oxidized form appear on the left of the equation.

An $E^{\ominus}$ value of $+0.76$ V means that iron (II) is a poorer reducing agent than hydrogen.

Table 12 gives a number of standard redox potentials. It must be remembered that these values refer to reactions in aqueous solution, at the temperature, pH and concentrations specified.

The importance of a standard oxidation–reduction potential is that it gives a measure of the tendency of an element to change from one oxidation state to another.

In Table 12, it is clear that the reducing power of the reduced form increases as the $E^{\ominus}$ value for a couple becomes more negative. Thus, it follows that:

(a) The reduced form of a couple (i.e., the species on the right of the equation) will reduce the oxidized form of another couple which has a less negative $E^{\ominus}$ value.
(b) The oxidized form of a couple (i.e., the species on the left of the equation) can oxidize the reduced form of any other couple which has a less positive $E^{\ominus}$ value.

Hence, tin (II) ion ($E^{\ominus}$ for Sn^{4+}/Sn^{2+} is $+0.15$ V) can reduce iron (III) to iron (II) (for the couple Fe^{3+}/Fe^{2+}, $E^{\ominus}$ is $+0.76$ V).
(c) In general, if the $E^{\ominus}$ values for two couples differ by more than

71

TABLE 12 *Standard oxidation–reduction potentials*

| Half reaction | | $E^\ominus$ | |
Oxidized form	Reduced form		
$K^+ + e^- = K$		-2.92 V	
$Ba^{2+} + 2e^- = Ba$		-2.90 V	
$Ca^{2+} + 2e^- = Ca$		-2.87 V	
$Na^+ + e^- = Na$		-2.71 V	
$Mg^{2+} + 2e^- = Mg$		-2.37 V	Reduced form
$Al^{3+} + 3e^- = Al$		-1.67 V	becoming a stronger
$Mn^{2+} + 2e^- = Mn$		-1.18 V	reducing agent
$Zn^{2+} + 2e^- = Zn$		-0.76 V	
$Cr^{3+} + 3e^- = Cr$		-0.74 V	
$Fe^{2+} + 2e^- = Fe$		-0.44 V	
$Sn^{2+} + 2e^- = Sn$		-0.14 V	
$H^+ + e^- = \frac{1}{2}H_2$		zero	
$Sn^{4+} + 2e^- = Sn^{2+}$		$+0.15$ V	
$Cu^{2+} + 2e^- = Cu$		$+0.34$ V	
$\frac{1}{2}I_2 + e^- = I^-$		$+0.54$ V	
$O_2 + 2H^+ + 2e^- = H_2O_2$		$+0.68$ V	
$Fe^{3+} + e^- = Fe^{2+}$		$+0.76$ V	
$Ag^+ + e^- = Ag$		$+0.80$ V	Oxidized form
$\frac{1}{2}Br_2 + e^- = Br^-$		$+1.07$ V	becoming a stronger
$\left.\begin{array}{l}Cr_2O_7^{2-} + 14H^+ \\ \quad + 6e^-\end{array}\right\} = \left\{\begin{array}{l}2Cr^{3+} \\ +7H_2O\end{array}\right.$		$+1.33$ V	oxidizing agent
$\frac{1}{2}Cl_2 + e^- = Cl^-$		$+1.36$ V	
$\left.\begin{array}{l}MnO_4^- + 8H^+ \\ \quad + 5e^-\end{array}\right\} = \left\{\begin{array}{l}Mn^{2+} \\ +4H_2O\end{array}\right.$		$+1.52$ V	
$\frac{1}{2}F_2 + e^- = F^-$		$+2.80$ V	

0.4 V, the reaction usually goes to completion. When the $E^\ominus$ differences are less than this, an equilibrium reaction is usually set up.

(d) Oxidation–reduction potentials measure *relative* tendencies for an element to change its oxidation state. Thus tin (II) can act as a reducing agent, while dichromate (VI) ion is an oxidizing agent, although they both have positive $E^\ominus$ values.

(e) In all the calculations, no conclusions are reached in regard to the speed at which the reaction will take place. This depends on other factors, and it is possible that a reaction thought feasible from $E^\ominus$ values, may not in fact, proceed at a measurable rate.

(f) $E^\ominus$ values refer to standard conditions, that is pH zero, and equimolar concentrations of oxidized and reduced forms.

(g) Under other conditions, and especially at a different pH value, the oxidation–reduction potential changes (and the reaction may take a different course).

Redox potentials determined under conditions other than standard *are not standard potentials*, and are designated by E instead of $E^\ominus$.

Oxidation State Charts

These are useful in displaying and relating the various oxidation states shown by an element. The oxidation state chart for chromium is shown in Figure 40. The arrows point from the oxidized to the

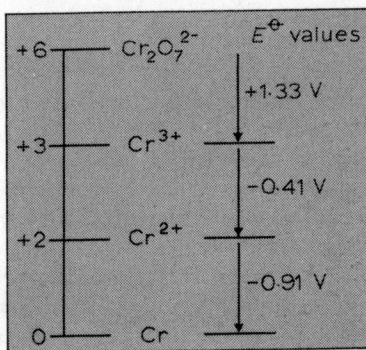

Fig. 40 Oxidation state chart for chromium

reduced form, in correspondence with the convention for writing redox equations. The $E^\ominus$ values link the three oxidation states shown. In this chart, dichromate (VI) ion (at pH zero) is a better oxidizing agent than hydrogen ion, while both chromium (II) and chromium metal are better reducing agents than hydrogen. This indicates that chromium (III) is the preferred oxidation state under the standard conditions in aqueous solution.

6 Section 1: General Chemistry of the Metallic Elements

The metallic elements are those which lie to the left of, and below, the diagonal line running from boron to polonium as shown in the Periodic Table (Table 1, p. 7).

In addition:

1 Elements lying close to the dividing line may show both metallic and non-metallic properties, so the division between the two sections is not sharply defined.
2 The most electropositive metals are those which lie on the extreme left of the Periodic Table, that is immediately following a noble gas.
3 The number of metallic elements increases in successive horizontal rows, especially in the fourth period which includes the first series of transition metals, and in the sixth period when the lanthanides appear.
4 Within a given group, the electropositive character of metals increases with increasing atomic number (although the reverse occurs among the transition metals). Consequently, in some groups, the early members may be typically non-metallic, while the later members are metals (e.g., carbon and silicon are typically non-metallic, while the later members of Group IV, tin and lead, are metals).
5 The properties of elements lying on a diagonal line in the Periodic Table often show close resemblances. For example, magnesium is less reactive than calcium (since it has a lower atomic number) and also less reactive than sodium. Thus, sodium and calcium might be expected to be about equally reactive—which is reflected in their standard electrode potentials (calcium $-2\cdot87$ V, sodium $-2\cdot71$ V):

$$\text{sodium} \longleftarrow \text{magnesium}$$
$$\downarrow$$
$$\dashrightarrow \text{calcium}$$

This is a particular example of the *diagonal relationships* that hold among elements. (Note that the line dividing metals and non-metals is also a diagonal.)

The ease of losing an electron (which is an indication of the reactivity of a metal) depends on:

(a) The size of the metal atom. The electron lost on ionization is further from the nucleus in the case of a large atom, and is thus less strongly bound that the electron lost by a small atom. Hence, the reactivity of metals within a group increases with increasing atomic size, that is increasing atomic number.

(b) The total number of electrons lost on ionization. A Group I metal forms an ion by losing one electron, while a Group II metal loses two electrons. The loss of two electrons is less easily accomplished, so that the Group II metals are less reactive than the corresponding metals in Group I (e.g., calcium is less reactive than potassium) but more reactive than the corresponding metals in Group III.

A combination of these two factors gives the theoretical explanation of the observed diagonal relationships.

The practical determination of the ease of losing electrons (i.e., the reactivity) may be obtained by measuring:

(a) Ionization energies. These refer to isolated gaseous metal atoms and do not apply to the aqueous system in which much of our chemistry is carried out.

(b) Standard electrode potentials (giving rise to the electrochemical series, as explained on p. 69).

The Importance of the Electrochemical Series

The displacement of hydrogen from strong acids All metals standing above hydrogen in the series (see Table 11, p. 69) can form ions more readily than hydrogen. Hence they should be capable of displacing hydrogen ion (and producing hydrogen gas) from solution. The ease with which this is accomplished decreases as the electrode potential steadily becomes less negative, so that metals close to hydrogen in the series react very slowly, while those standing below hydrogen do not displace hydrogen at all. For example,

$$Zn + 2H^+ = Zn^{2+} + H_2$$

Note that nitric acid attacks nearly all metals, although hydrogen is not generally among the products, since it is oxidized to water by nitric acid in all but very dilute solutions.

Reaction of the metals with water As the electrode potentials for the various metals become progressively more negative, the metals become capable of liberating hydrogen from solutions containing successively lower concentrations of hydrogen ion. Thus, metals at the top of the electrochemical series displace hydrogen from cold water, while those lower down displace hydrogen from hot water or steam. For example, potassium, barium, calcium and sodium react with cold water:

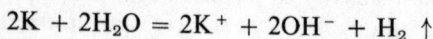

$$2K + 2H_2O = 2K^+ + 2OH^- + H_2 \uparrow$$

Magnesium, when finely divided, reacts slowly with cold water; it reacts more rapidly with hot water and the metal burns in steam.

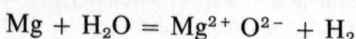

$$Mg + H_2O = Mg^{2+} O^{2-} + H_2$$

Red-hot aluminium and zinc react with steam

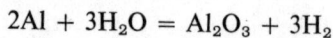

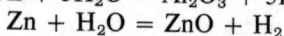

$$2Al + 3H_2O = Al_2O_3 + 3H_2$$
$$Zn + H_2O = ZnO + H_2$$

The reaction between red-hot iron and steam is reversible.

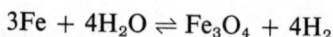

$$3Fe + 4H_2O \rightleftharpoons Fe_3O_4 + 4H_2$$

Metals appearing below iron in the electrochemical series show very little reaction towards water.

The e.m.f. of a chemical cell

(a) A cell capable of producing an electric current is made by coupling two dissimilar metal electrodes, immersed in a suitable electrolyte.

(b) Provided the temperature is 25°C and the electrolyte concentration is one mole per litre of each metal ion, the e.m.f. of the cell is given by the difference between the two standard electrode potentials. For example, the Daniell cell comprises a copper electrode coupled with a zinc electrode. Under standard conditions, the e.m.f. of this cell is

$$0.34 - (-0.76) = 1.1 \text{ V}$$

The products of electrolysis The discharge process which liberates cations at the cathode is

$$M^{n+} + ne^- = M$$

Thus metals at the top of the electrochemical series which lose electrons easily are reluctant to accept them. Consequently, the ease of discharge of the metal ions is greatest with metals at the bottom of the series, and it becomes more difficult to discharge cations as the standard electrode potential becomes progressively more negative. Hence, electrolysis of a solution containing copper and iron ions deposits copper first.

Displacement of one metal by another   A metal will displace another metal which stands below it in the electrochemical series, from a solution of any of its salts. For example,

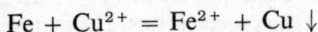

$$Fe + Cu^{2+} = Fe^{2+} + Cu \downarrow$$

Reduction of metal ores   Methods are dependent on the position in the electrochemical series.

1 Reduction involves the *gain of electrons* by a species (see p. 65).
2 A metal is combined in an ore, and is therefore present as a cation. Reduction of the ore involves the process:

$$M^{n+} + ne^- = M$$

3 The process becomes progressively easier for metals lower in the electrochemical series (see also 'The products of electrolysis' on p. 76).
4 Metals low down in the series are usually reduced by heating the ore under controlled conditions. This is termed *self-reduction*.

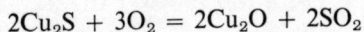

$$2Cu_2S + 3O_2 = 2Cu_2O + 2SO_2$$

By roasting in air, the sulphide ore is converted into the oxide. Before this process is complete, the air supply is cut off, and further heating brings about a reaction between unchanged sulphide and the oxide:

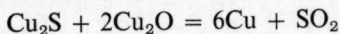

$$Cu_2S + 2Cu_2O = 6Cu + SO_2$$

A similar process is used for lead.
5 Cations of the metals high in the electrochemical series, (potassium to aluminium) accept electrons with difficulty. For reduction of these ores, electrolysis of a fused salt, is necessary.
   For example, sodium chloride (mixed with calcium chloride

to lower the melting point) is fused and electrolysed in the Down's Cell. At the cathode,

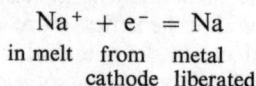

$$Na^+ + e^- = Na$$

in melt   from   metal
cathode liberated

At the anode,

$$2Cl^- = Cl_2 + 2e^-$$

Note: The ion at the cathode accepts electrons and is thus reduced, while the ion at the anode gives up electrons and so is oxidized. Consequently, electrolysis is a method of oxidation and reduction.

6 Metals in the centre of the electrochemical series are reduced from their ores by chemical reduction. Common reducing agents are carbon and carbon monoxide. The following discussion is restricted to oxide ores, since the majority of metals extracted by chemical reduction are found as oxide ores.

It was shown on pp. 59–62 that the tendency for a reaction to take place was determined by:

(a) the change in free energy ΔG,
(b) kinetic factors.

Taking the first factor, the free energy change for the formation of a metal oxide, for example zinc oxide, is:

$$Zn + \tfrac{1}{2}O_2 = ZnO; \quad \Delta G^\ominus = -627 \text{ kJ}$$

($\Delta G^\ominus$ refers to values obtained under standard conditions—one of which is that the temperature is fixed at 25°C.) Thus, to separate 1 mole of zinc oxide into its elements, 627 kJ of free energy must be supplied, and this is the function of the reducing agent.

Using carbon as a reducing agent, the corresponding free energy change is:

$$C + \tfrac{1}{2}O_2 = CO; \quad \Delta G^\ominus = -230 \text{ kJ} \quad \text{(again at 25°C)}$$

Under standard conditions, when carbon reacts with oxygen to form carbon monoxide, 230 kJ of free energy is liberated, and this falls well short of the 627 kJ required. However, as the temperature rises, the free energy changes alter:

78

$$C + \tfrac{1}{2}O_2 = CO; \quad \Delta G = -439 \text{ kJ at } 1000°C \qquad \text{(i)}$$
$$Zn + \tfrac{1}{2}O_2 = ZnO; \quad \Delta G = -377 \text{ kJ at } 1000°C \qquad \text{(ii)}$$

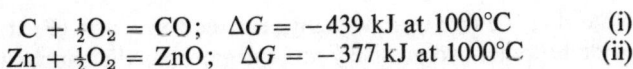

Subtracting (ii) from (i),

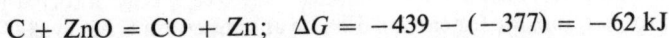

$$C + ZnO = CO + Zn; \quad \Delta G = -439 - (-377) = -62 \text{ kJ}$$

At 1000°C, the reduction of zinc oxide by carbon is possible, since the free energy change for the reaction is now negative.

The second factor—the speed at which the reaction takes place—is determined by the activation energy and the mechanism of the reaction, and this aspect is discussed in Part 3 of this series, *Physical Chemistry*.

The variation of free energy changes with temperature is an important factor in connection with these extraction processes. The data, shown graphically in what is known as an *Ellingham diagram*, are shown in Figure 41. The breaks in the individual graphs for the metal oxides are due to changes in the physical state of one of the species shown in the equation for oxide formation. In all the cases shown in Figure 41, the breaks correspond to the point at which the metal melts (points *M* on the chart) or boils (points *B*).

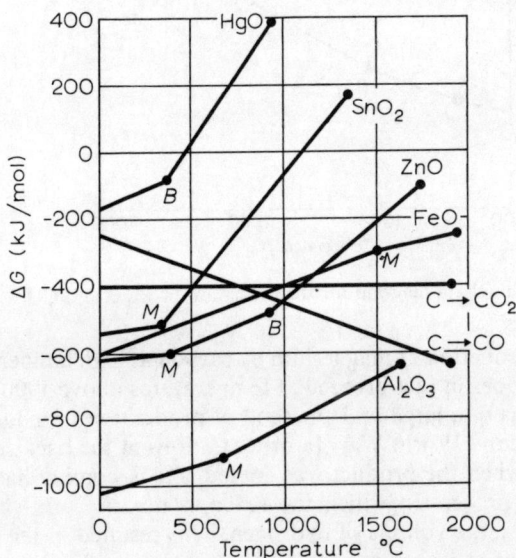

Fig. 41 Ellingham diagram

Note that the free energy change for the formation of carbon monoxide becomes progressively more negative as the temperature rises. This factor is particularly significant in producing the strong reducing power of carbon at higher temperatures.

Carbon will reduce the metal oxide, if the free energy of formation of carbon monoxide (or carbon dioxide at temperatures below 800°C) is more negative than the free energy of formation of the metal oxide. Thus, at 1000°C, carbon reduces mercury (II) oxide, tin (IV) oxide, zinc and iron (II) oxides (producing carbon monoxide and the metal), but aluminium oxide is not reduced. The free energy graph for aluminium oxide lies well below the graphs for other metals included on the chart. This explains why aluminium may be used to reduce other metals from their oxides (the Thermite process).

The Ellingham diagram (Figure 42) for the formation of magnesium oxide and carbon monoxide shows that it is feasible to

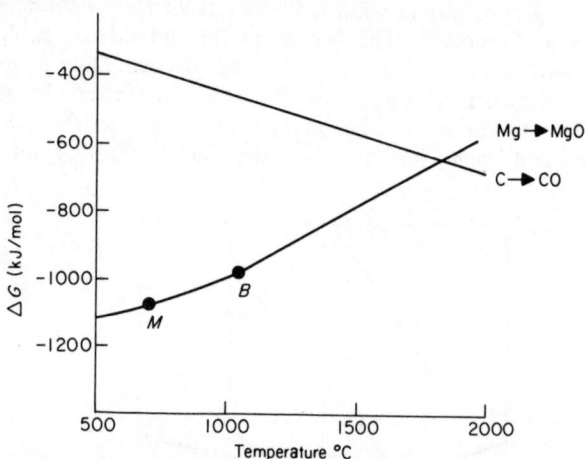

Fig. 42 Ellingham diagram for the formation of MgO and CO

consider the reduction of magnesium by carbon at high temperatures. The reaction does in fact proceed at temperatures above 1900°C and it was the basis of a large-scale method of production of magnesium during the Second World War. In order to prevent the back reaction taking place when the product was cooled, shock cooling had to be employed—a process in which the reaction mixture was suddenly diluted with a large volume of hydrogen. This resulted in the formation of very fine magnesium powder which proved difficult to handle on account of the ease with which it caught fire.

The corrosion of galvanized iron and tinplate Metals high in the electrochemical series react easily, and are not in general used to withstand corrosive attack. Metals low in the series such as lead, copper and platinum are often used for such purposes. Other factors have also to be considered:

(a) The formation of a stable protective film, as with zinc and aluminium which form stable oxide films.
(b) Contact with a second metal.

When two dissimilar metals are in contact, and the junction is wetted by an electrolyte solution, a cell is set up. In such cases the metal higher in the electrochemical series dissolves. Hence, corrosive attack often occurs at the junction of two dissimilar metals. This factor also explains why very pure metals are more resistant to acid attack than a metal which contains some impurities. Iron is often coated with a layer of tin (tinplate) or zinc (galvanizing) to improve its corrosion resistance. Should the film of protecting metal be damaged, and the iron exposed, zinc is dissolved first in the case of galvanized iron, while iron is attacked first in the case of tinplate. Thus galvanized iron does not rust when small areas of iron are exposed, whereas tinplate rusts rapidly in similar circumstances.

THE EXTRACTION AND USES OF THE COMMONER METALS

Sodium

This metal is extracted from sodium chloride ore by electrolysis of a fused mixture of sodium and calcium chlorides (the fusion temperature being lower for such a mixture). The cell used is the Downs cell (Figure 43). It consists of a circular steel shell lined with firebrick.
    At the cathode:

$$Na^+ + e^- = Na$$

The molten sodium is led to the offtake by the cathode shield.
    At the anode:

$$Cl^- = \tfrac{1}{2}Cl_2 + e^-$$

Uses   (a) As a heat exchange fluid in nuclear power plants.

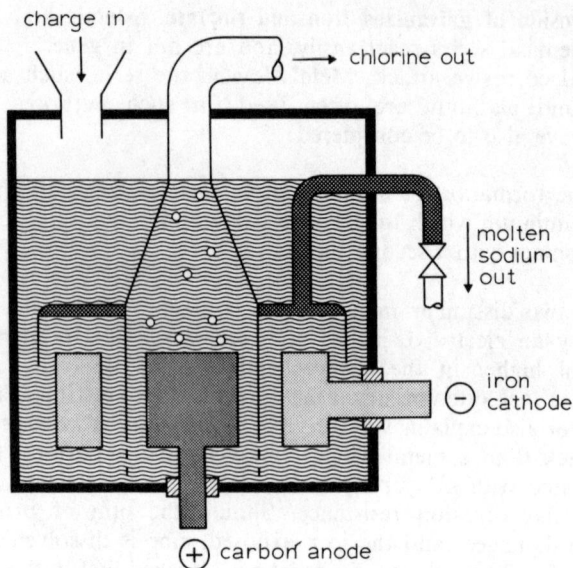

Fig. 43 Downs cell

(b) In the extraction of rarer and more expensive metals, for example, titanium, by the reaction:

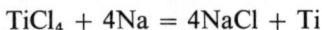

$$TiCl_4 + 4Na = 4NaCl + Ti$$

(c) In the preparation of petrol additives.
(d) As a reducing agent (sodium and alcohol, or sodium amalgam).
(e) In discharge (sodium vapour) lamps.

Magnesium

The carbonate ore is converted to the oxide, which is then pelleted with carbon, and treated with chlorine:

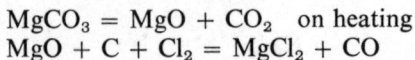

$$MgCO_3 = MgO + CO_2 \quad \text{on heating}$$
$$MgO + C + Cl_2 = MgCl_2 + CO$$

Magnesium is extracted by electrolysis of the fused magnesium chloride produced in this reaction.

At the cathode: $Mg^{2+} + 2e^- = Mg$

The magnesium is removed by scooping off the molten metal from the cathode compartment from time to time.

At the anode: $2Cl^- = Cl_2 + 2e^-$

Uses (a) Mainly for lightweight structural alloys.
(b) In flares and flash powders.
(c) As a 'sacrificial' metal in cases of anodic corrosion.
(d) In organic chemistry—for example Grignard reagents.

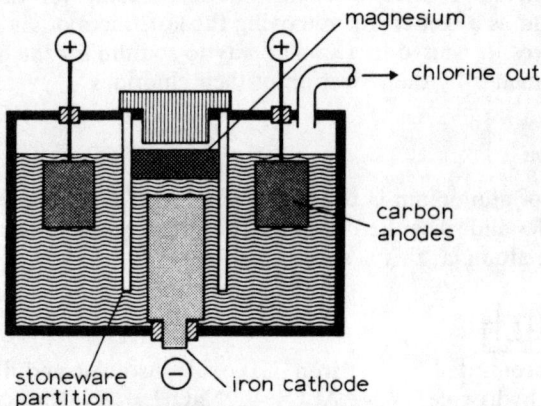

Fig. 44 Magnesium cell

Calcium

Fused calcium chloride (which is a by-product of the Solvay process) is electrolysed in a steel cell, with graphite wall linings. The cathode

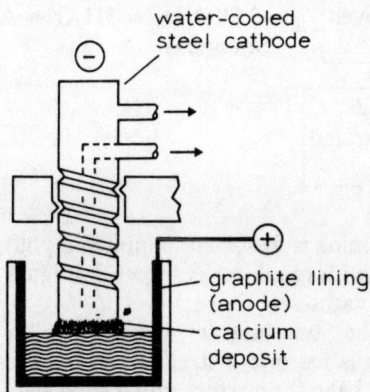

Fig. 45 Calcium cell

is a water-cooled iron rod which is gradually withdrawn as the calcium deposit builds up.

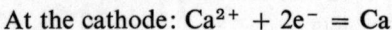

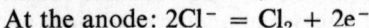

At the cathode: $Ca^{2+} + 2e^- = Ca$
At the anode: $2Cl^- = Cl_2 + 2e^-$

Uses Calcium is used in small amounts mainly for deoxidizing metals and as a 'getter' for removing the last traces of air in thermionic valves. It is used in a similar way to sodium for the extraction of other metals by the reduction of their chlorides.

Aluminium

The ore of aluminium is bauxite, a mixture of aluminium and iron (III) oxides and silica. Before electrolysis, the ore has to be converted into pure alumina; this is effected as follows:

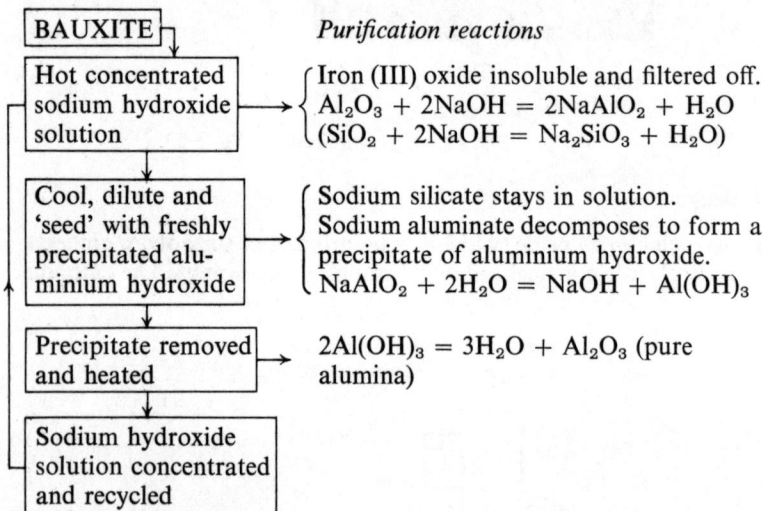

BAUXITE	*Purification reactions*
Hot concentrated sodium hydroxide solution	$\left\{\begin{array}{l}\text{Iron (III) oxide insoluble and filtered off.}\\ Al_2O_3 + 2NaOH = 2NaAlO_2 + H_2O\\ (SiO_2 + 2NaOH = Na_2SiO_3 + H_2O)\end{array}\right.$
Cool, dilute and 'seed' with freshly precipitated aluminium hydroxide	$\left\{\begin{array}{l}\text{Sodium silicate stays in solution.}\\ \text{Sodium aluminate decomposes to form a}\\ \text{precipitate of aluminium hydroxide.}\\ NaAlO_2 + 2H_2O = NaOH + Al(OH)_3\end{array}\right.$
Precipitate removed and heated	$2Al(OH)_3 = 3H_2O + Al_2O_3$ (pure alumina)
Sodium hydroxide solution concentrated and recycled	

Electrolysis Alumina is dissolved in molten cryolite (Na_3AlF_6). The cell is steel, lined with graphite, as shown in Figure 46. The graphite lining acts as the cathode.

The anodes are the Söderberg self-baking type, in which a mixture of coke and pitch is fed into a steel casing. The heat emitted by the cell is sufficient to bake the mixture into a solid electrode. The anodes are gradually consumed during the process, and they are continually lowered into the cell to within two or three inches of the molten

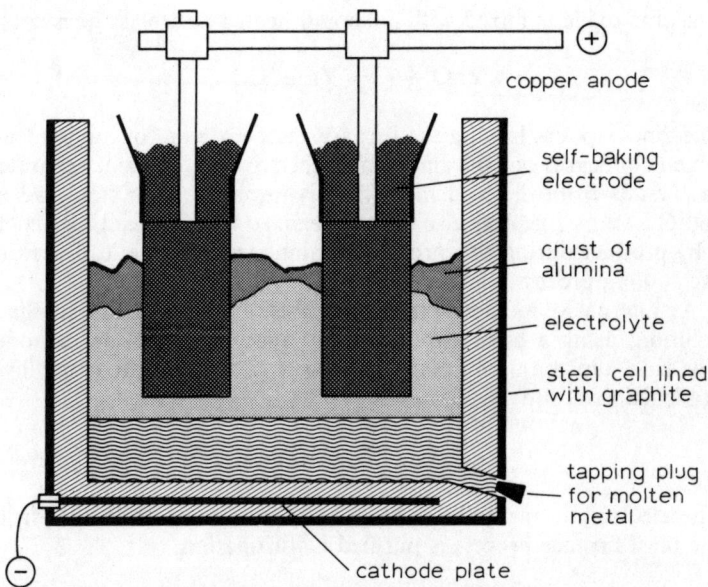

Fig. 46 Electrolytic extraction of aluminium

aluminium. The cell runs at 5 or 6 volts and at 32 000 amps. The discharge process is probably more involved than the equations given below suggest.

At the cathode: $2Al^{3+} + 6e^- = 2Al$
At the anode: $3O^{2-} = 1\frac{1}{2}O_2 + 6e^-$

At the temperature of operation, the oxygen burns away the carbon anode. From time to time, the alumina crust is broken and more alumina (together with a small amount of cryolite to make good any wastage) is added.

Uses (a) In lightweight yet strong alloys, especially in the aircraft industry.
  (b) Electrical cables and insulators.
  (c) Household utensils.
  (d) As a packaging material.

Zinc
The sulphide ore is roasted in air to form the oxide:

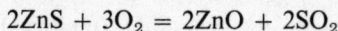

$$2ZnS + 3O_2 = 2ZnO + 2SO_2$$

85

The zinc oxide is mixed with coke and heated in a blast furnace:

$$ZnO + C = Zn + CO$$

The zinc vapours leaving the furnace meet a stream of molten lead which condenses out the zinc. On further cooling, the zinc separates as crystals from the lead melt (zinc is insoluble in molten lead at 450°C). Many zinc ores also contain lead, so that any lead formed as a by-product during this process accumulates with the lead used in the cooling process.

An alternative method is the electrolysis of a purified zinc sulphate solution, using a high current density and an aluminium cathode. The zinc sulphate used for this process is obtained by roasting low-grade sulphide ore:

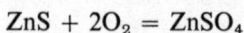

$$ZnS + 2O_2 = ZnSO_4$$

The electrolytic method produces very pure zinc, while zinc from the blast furnace process is purified by distillation.

Uses   (a) As a protective coating for iron (galvanizing, sherardizing, etc.).
(b) In alloys (brass is a copper–zinc alloy).
(c) In dry batteries.

Iron

The reduction of the oxide ores of iron is accomplished in a blast furnace, the charge consisting of coke, limestone (to form a slag with the silicate material included in the ore) and iron oxide ore. The reactions are as follows:

Upper zone of furnace; temperature 400°–600°C:

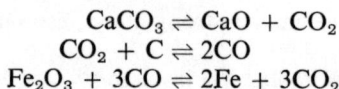

$$CaCO_3 \rightleftharpoons CaO + CO_2$$
$$CO_2 + C \rightleftharpoons 2CO$$
$$Fe_2O_3 + 3CO \rightleftharpoons 2Fe + 3CO_2$$

In the upper zone of the furnace, the iron is formed as a spongy mass which readily picks up sulphur from the fuel.

Central zone of the furnace; temperature 600°–1000°C:

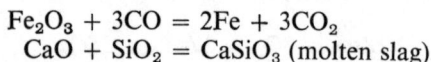

$$Fe_2O_3 + 3CO = 2Fe + 3CO_2$$
$$CaO + SiO_2 = CaSiO_3 \text{ (molten slag)}$$

Very hot zone (just above the tuyères); temperature about 1200°C:

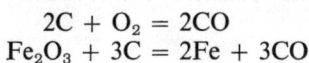

$$2C + O_2 = 2CO$$
$$Fe_2O_3 + 3C = 2Fe + 3CO$$

Phosphorus and silicon are both formed in this zone by the reduction of phosphate and silicate impurities in the ore. The iron produced contains sulphur, phosphorus, silicon and carbon as the chief impurities. The blast used in modern furnaces consists of pure oxygen and fuel oil, and the presence of the oil reduces the quantity of coke needed in the initial charge. The waste gases are rich in carbon monoxide, and are useful as a fuel in preheating the blast.

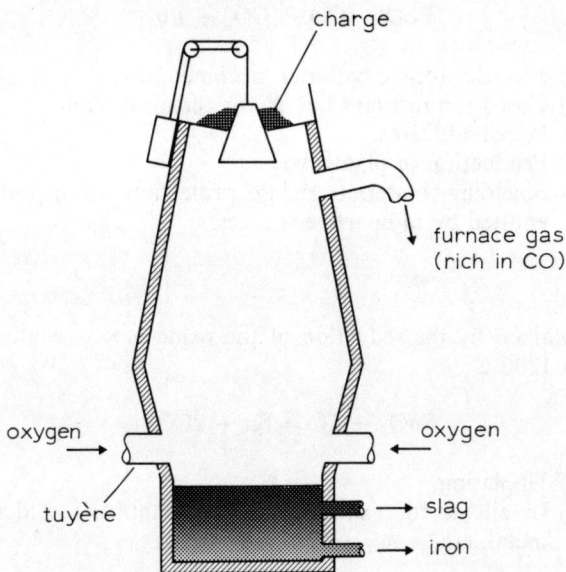

Fig. 47 Blast furnace

Uses The product of the blast furnace is pig iron, some of which is used for making cast iron, while most of the remainder is converted into steel.

Lead

A great deal of the output of lead comes from the recovery of scrap metal.

The extraction of lead from the sulphide ore (which is concentrated and separated from gangue by froth flotation) takes place in two stages. The ore is first roasted in air, under carefully controlled conditions:

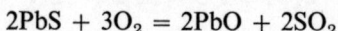

$$2PbS + 3O_2 = 2PbO + 2SO_2$$

The air supply is then cut off and the temperature is raised:

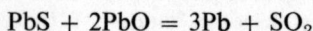

$$PbS + 2PbO = 3Pb + SO_2$$

A more modern method involves roasting the ore to the oxide as above; this is then mixed with coke and reduced in a blast furnace:

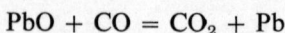

$$PbO + CO = CO_2 + Pb$$

Uses (a) Electric storage batteries (accumulators).
(b) Corrosion resistant linings for chemical plant.
(c) Petrol additives.
(d) Production of pigments.
(e) Shielding material giving protection from radiations emitted by radioactive sources.

Tin

Tin is obtained by the reduction of the oxide ore by heating with carbon at 1200°C:

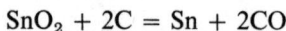

$$SnO_2 + 2C = Sn + 2CO$$

Uses (a) Tinplating.
(b) In alloys, for example, bronze, plumber's solder, type metal, etc.

Copper

Copper is extracted from the sulphide ore by a process of *self-reduction*, the stages being as follows:

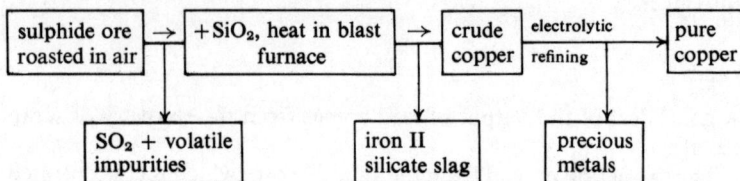

The ore is $CuS.FeS$. The reactions are as follows:

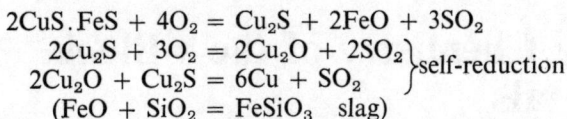

$$2CuS.FeS + 4O_2 = Cu_2S + 2FeO + 3SO_2$$
$$\left.\begin{array}{l} 2Cu_2S + 3O_2 = 2Cu_2O + 2SO_2 \\ 2Cu_2O + Cu_2S = 6Cu + SO_2 \end{array}\right\}\text{self-reduction}$$
$$(FeO + SiO_2 = FeSiO_3 \quad \text{slag})$$

The crude copper produced in the blast furnace is refined by electrolysis, using a thin sheet of pure copper as the anode, a block of crude copper as cathode, and acidified copper sulphate as the electrolyte. It is usual to feed into the extraction process, quantities of low-grade precious metal ores. These metals are recovered from the 'anode slime' deposited during the refining process.

Uses (a) Pipework, utensils and chemical plant.
        (b) Electrical wires and cables.
        (c) Alloys, such as brass and bronze.

Mercury

The sulphide ore decomposes on strong heating, and the mercury produced may be purified by distillation.

Uses (a) High density liquid in scientific instruments.
        (b) Electrically conducting liquid used in contacts and relays, etc.
        (c) Discharge lamps.
        (d) Organic mercury compounds for use as fungicides and antiseptics.

Silver

The metal is obtained from the sulphide ore by way of complex cyanide formation:

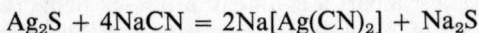

$$Ag_2S + 4NaCN = 2Na[Ag(CN)_2] + Na_2S$$

The soluble cyanide complex is decomposed by zinc in alkaline solution:

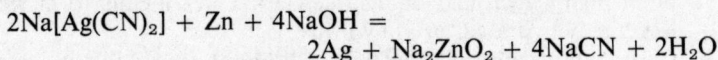

$$2Na[Ag(CN)_2] + Zn + 4NaOH =$$
$$2Ag + Na_2ZnO_2 + 4NaCN + 2H_2O$$

Uses (a) Silverware and silver plating.
        (b) Production of silver compounds used in photography.
        (c) Silvering mirrors (vacuum deposition of aluminium is a modern alternative to this process).

7 The Chemistry of the s-Block Metals

GROUP 1A: THE ALKALI METALS

General Properties

1 Soft silvery-white metals, of low density. Easily cut, and are malleable and ductile. They have low melting points.
These properties indicate that the metallic bonding system is rather weak since:
  (a) The atoms of the alkali metals are large, and thus the structural units in the metal crystal are bulky; hence the crystal has a low density.
  (b) Only one valence electron per atom enters the metallic bond. The weak metallic bond results from a combination of these two factors; that is, a bulky structure is held by the minimum number of binding electrons.

2 Good conductors of heat and electricity.

3 Highly reactive metals (refer to their position in the electro-chemical series). They react vigorously with water.

4 Form colourless, unipositive cations.

5 The oxides are basic and dissolve in water to produce a strongly alkaline solution.

6 Almost all their salts are freely soluble in water.

7 React with hydrogen on heating, and under pressure to form solid 'salt-like' hydrides, M^+H^-.

8 All the metals impart a colour to a bunsen flame (lithium, red; sodium, yellow; potassium, violet; rubidium, bluish red; caesium, blue).

9 Salts of metals high in the group (i.e., lithium and sodium) are often highly hydrated or deliquescent. Later metals form salts less highly hydrated, or anhydrous,

10 Carbonates (with the exception of lithium) are soluble in water, and stable to heat.

In many of its properties, lithium stands apart from the rest of the Group I metals. These differences will be outlined later (p. 97).

Typical Metal—Sodium

Ore: sodium chloride

$E^\ominus$ **value:** $-2 \cdot 71$ V

Reaction of metal with:

(a) *Air* (on heating),
$$2Na + O_2 = Na_2O_2$$
(sodium peroxide formed with excess air);
$$4Na + O_2 = 2Na_2O$$
(sodium monoxide formed with excess metal).

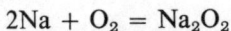

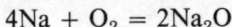

(b) *Water* (in the cold),
$$2Na + 2H_2O = 2NaOH + H_2$$

(c) *Ammonia gas* (on heating),
$$2Na + 2NH_3 = 2NaNH_2 + H_2$$
(sodamide is formed).

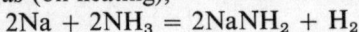

Sodium dissolves in liquid ammonia to give a blue solution, while no hydrogen is evolved. These solutions are good conductors of electricity and are strongly reducing.

(d) *Hydrogen* (on heating under pressure),
$$2Na + H_2 = 2NaH$$
Electrolysis of the product, sodium hydride, in a suitable melt, produces hydrogen at the anode, thus showing the presence of the hydride (H^-) ion.

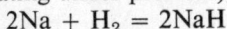

Outline of the Main Reactions of Sodium

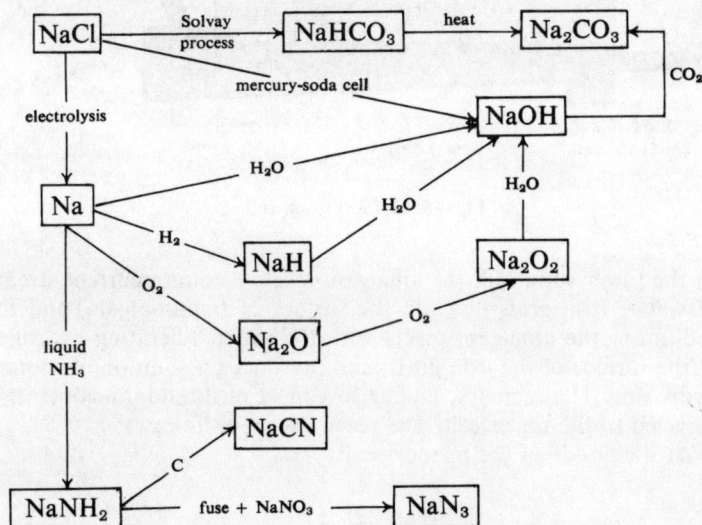

The Manufacture of Some Important Sodium Compounds

Sodium hydroxide: the mercury-soda cell The cell is shown diagrammatically in Figure 48. In the upper cell, the mercury cell, chlorine is given off at the anode. At the flowing mercury cathode, sodium ions are discharged (since hydrogen acquires a large overpotential at a mercury surface) and form an amalgam with the mercury.

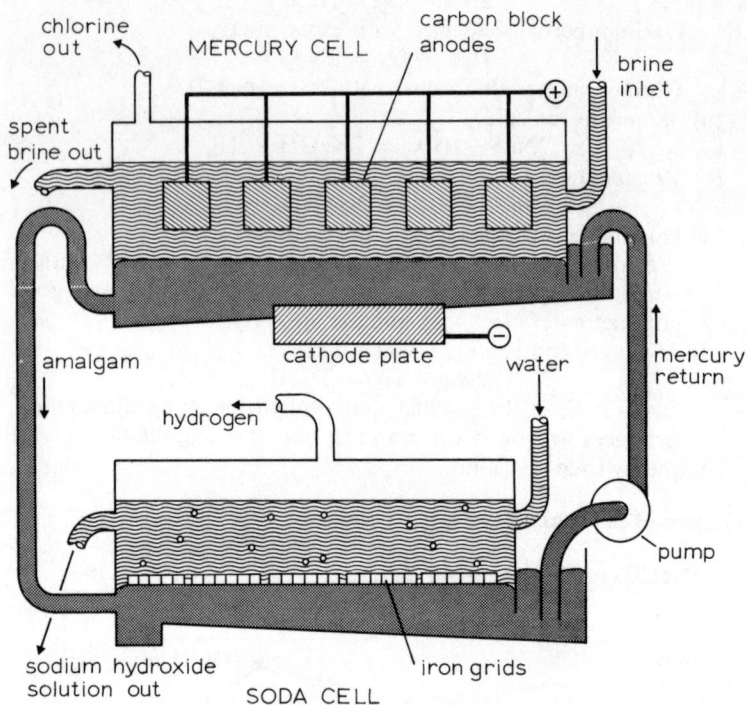

Fig. 48 Mercury-soda cell

In the lower soda cell, the amalgam meets a countercurrent stream of water. Iron grids float on the surface of the amalgam, and the sodium in the amalgam reacts with the water, liberating hydrogen at the surface of the iron grids, and producing a solution of sodium hydroxide. The mercury, having lost most of its sodium content, is recycled to the upper cell. The reactions are as follows:

At the anode of the mercury cell:

$$Cl^- = \tfrac{1}{2}Cl_2 + e^-$$

92

At the cathode (ions present, Na^+ and H^+), Na^+ discharged preferentially:

$$Na^+ + e^- = Na \text{ (amalgam)}$$

In the soda cell:
$$2Na \text{ (amalgam)} + 2H_2O = 2NaOH + H_2$$

Sodium hydrogen carbonate: the Solvay process This is shown diagrammatically in Figure 49. In the ammonia absorption tower, ammonia and brine are mixed to form *ammoniated brine:*

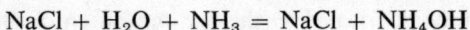

$$NaCl + H_2O + NH_3 = NaCl + NH_4OH$$

The ammoniated brine is passed down the carbonating or Solvay tower. This tower is fitted with baffles to ensure thorough mixing of the carbon dioxide and the ammoniated brine. Cooling coils are

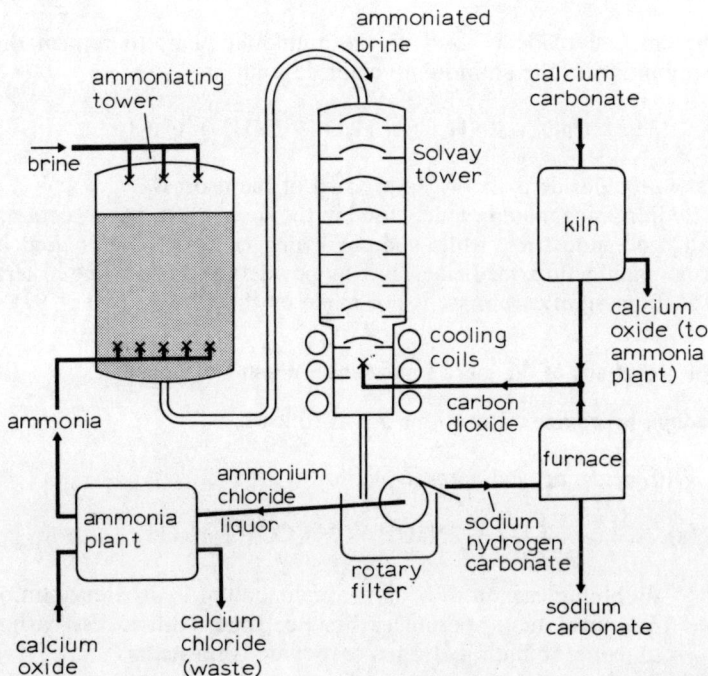

Fig. 49 Solvay process

93

fitted near the base of the tower, since sodium hydrogen carbonate is not very soluble in the cold. The following reactions take place:

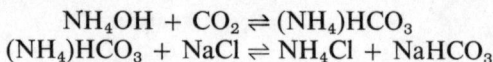

$$NH_4OH + CO_2 \rightleftharpoons (NH_4)HCO_3$$
$$(NH_4)HCO_3 + NaCl \rightleftharpoons NH_4Cl + NaHCO_3$$

The sodium hydrogen carbonate slurry is fed to the rotary vacuum filter, and the solid product is then heated to produce sodium carbonate:

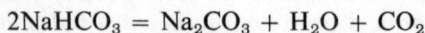

$$2NaHCO_3 = Na_2CO_3 + H_2O + CO_2$$

The carbon dioxide formed by the decomposition of sodium hydrogen carbonate is mixed with an equal quantity of carbon dioxide (made by heating calcium carbonate) and passed back to the Solvay tower:

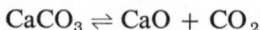

$$CaCO_3 \rightleftharpoons CaO + CO_2$$

The calcium oxide is used in the ammonia plant to reform the ammonia from the ammonium chloride solution:

$$CaO + 2NH_4Cl = H_2O + 2NH_3 + CaCl_2$$

Calcium chloride is the waste product of the process.

Sodium carbonate is widely used in the soap, glass, paper, ceramic and food industries, while sodium hydrogen carbonate is used in drug manufacture, medicines, baking powders and fire extinguishers. N.B. Potassium carbonate is not made by this process (see p. 97).

The Reactions of the more Important Sodium Compounds

Sodium hydroxide Reactions are as follows:

1 With acids, or acid gases:

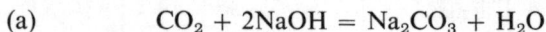

(a) $$CO_2 + 2NaOH = Na_2CO_3 + H_2O$$

Bubbling carbon dioxide through sodium hydroxide solution produces first, sodium carbonate; then, with excess carbon dioxide, sodium hydrogen carbonate is formed:

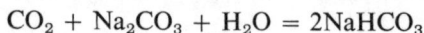

$$CO_2 + Na_2CO_3 + H_2O = 2NaHCO_3$$

(Sulphur dioxide reacts in a similar way.)

94

(b) $$HCl + NaOH = NaCl + H_2O$$

(c) $$H_2S + 2NaOH = Na_2S + 2H_2O$$

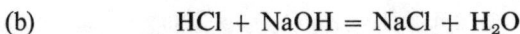

on first bubbling the gas through sodium hydroxide solution, then:

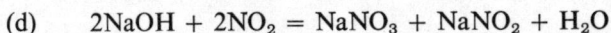

$$Na_2S + H_2S = 2NaHS$$

(d) $$2NaOH + 2NO_2 = NaNO_3 + NaNO_2 + H_2O$$

(e) When chlorine is bubbled through an excess of cold, dilute sodium hydroxide solution: sodium chlorate (I) and sodium chloride are formed.

$$Cl_2 + 2NaOH = NaCl + H_2O + NaOCl$$

Electrolysis of cold, dilute, well-stirred brine also leads to the formation of sodium chlorate (I).

When excess chlorine is passed through hot, concentrated sodium hydroxide solution, sodium chlorate (V) is formed:

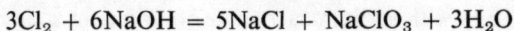

$$3Cl_2 + 6NaOH = 5NaCl + NaClO_3 + 3H_2O$$

2 With elements:

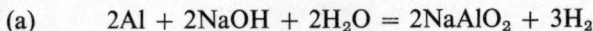

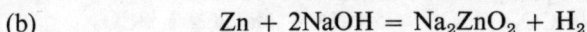

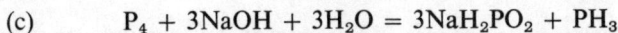

(a) $$2Al + 2NaOH + 2H_2O = 2NaAlO_2 + 3H_2$$

(b) $$Zn + 2NaOH = Na_2ZnO_2 + H_2$$

(c) $$P_4 + 3NaOH + 3H_2O = 3NaH_2PO_2 + PH_3$$

Yellow phosphorus and sodium hydroxide give phosphine on gentle heating.

3 With ammonium salts, or derivatives of ammonia:

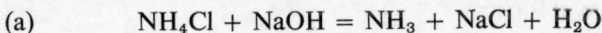

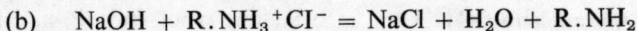

(a) $$NH_4Cl + NaOH = NH_3 + NaCl + H_2O$$

(b) $$NaOH + R.NH_3^+Cl^- = NaCl + H_2O + R.NH_2$$

4 With solutions of metal salts:

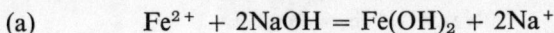

(a) $$Fe^{2+} + 2NaOH = Fe(OH)_2 + 2Na^+$$

95

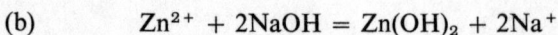

(b) $$Zn^{2+} + 2NaOH = Zn(OH)_2 + 2Na^+$$

Then,

$$Zn(OH)_2 + 2OH^- = ZnO_2^{2-} + 2H_2O$$

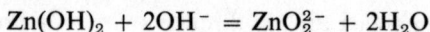

5 Hydrolysis of organic acid derivatives:

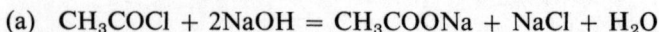

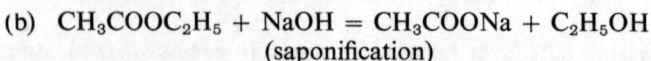

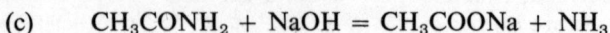

(a) $CH_3COCl + 2NaOH = CH_3COONa + NaCl + H_2O$

(b) $CH_3COOC_2H_5 + NaOH = CH_3COONa + C_2H_5OH$
(saponification)

(c) $CH_3CONH_2 + NaOH = CH_3COONa + NH_3$

Sodium carbonate Produced as large monoclinic crystals (Na_2CO_3. $10H_2O$) on cooling a hot concentrated solution. The crystals effloresce in air to $Na_2CO_3.H_2O$. The anhydrous material decomposes slightly on heating:

$$Na_2CO_3 = Na_2O + CO_2$$

Chlorine reacts with sodium carbonate solutions in a similar way to sodium hydroxide solutions, forming chlorate (I) with cold, dilute solutions of sodium carbonate, and chlorate (V) with hot, concentrated solutions and with an excess of chlorine:

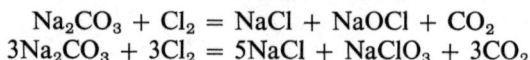

$$Na_2CO_3 + Cl_2 = NaCl + NaOCl + CO_2$$
$$3Na_2CO_3 + 3Cl_2 = 5NaCl + NaClO_3 + 3CO_2$$

Sodium hydrogen carbonate Decomposes on heating, or when placed in hot water:

$$2NaHCO_3 = Na_2CO_3 + H_2O + CO_2$$

The solubility is low in cold water, a saturated solution containing 13 g per litre at 0°C.

Sodium thiosulphate (VI) Used in bleaching wool and oil, and in photography. It is made either by the action of sulphur dioxide on sodium sulphide, or the reaction of sodium sulphite and sulphur:

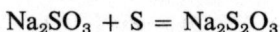

$$Na_2SO_3 + S = Na_2S_2O_3$$

Potassium Salts

In the main, the salts of potassium show a great resemblance to those of sodium. Notable exceptions are:

1 Potassium salts usually crystallize in the anhydrous state (cf. $Na_2SO_4 . 10H_2O$ and K_2SO_4).
2 Potassium chlorate (VII) ($KClO_4$) is sparingly soluble.
3 Potassium hydrogen carbonate is much more soluble than the corresponding sodium salt and, in view of this, the Solvay process cannot be applied to the production of potassium hydrogen carbonate.
4 Potassium carbonate is deliquescent. This is the reason why potassium hydroxide is often specified for the absorption of carbon dioxide, since the use of sodium hydroxide would lead to the formation of sodium carbonate, which in view of its efflorescent properties could lead to the blockage of a tube in an absorption train used intermittently.
5 Potassium forms a superoxide, KO_2.
6 Potassium salts are essential for plant life.

The Special Properties of Lithium

Many of the special properties of lithium exemplify Fajan's Rules (p. 54).

1 Reaction with water is slow.
2 Does not react with bromine in the cold.
3 Reacts directly with nitrogen to form lithium nitride (Li_3N).
4 Lithium hydride is the most stable hydride of Group I. This hydride has a high melting point (670°C), does not decompose readily on heating, and does not react with oxygen or dry gaseous hydrogen chloride.
5 On heating in air or oxygen, the monoxide only (Li_2O) is formed.
6 Both lithium hydroxide and lithium carbonate decompose to form the oxide on strong heating.
7 Forms a wide range of *organo-metallic* compounds, for example, lithium ethyl, empirical formula C_2H_5Li.
8 Most lithium salts are either deliquescent, or highly hydrated.
9 Lithium chloride, bromide and iodide are soluble in organic solvents.
10 Lithium sulphate does not form alums.

The Effect of the Charge-to-Volume Ratio of an Ion

The electrical field surrounding a small ion which carries a high charge, is intense. Such ions are said to be highly polarizing, and the polarizing power of an ion is related to the charge-to-volume ratio of the ion.

Although lithium forms a unipositive ion, the size of the lithium ion is small, and the electrical field surrounding it is intense. This factor explains:

(a) The thermal decomposition of lithium nitrate and carbonate. Both nitrate and carbonate ions are large planar ions, and they are soon distorted by a strong electrical field and hence decompose on heating in lithium salts.

(b) The deliquescent nature (and high degree of hydration) of lithium salts. The attraction of water molecules by the small lithium ion accounts for this effect, which gradually disappears with the increase in size of the ions of the later members of the group.

GROUP IIA: THE ALKALINE EARTH METALS

General Properties

1 Beryllium, and to some degree, magnesium, exhibit properties different from the remaining metals (calcium, strontium and barium), so that beryllium and magnesium are often excluded from the classification of alkaline earths.

2 The metals are uniformly divalent.

3 They form colourless cations.

4 They are soft metals, of low density (although they are more brittle and have a higher density than the corresponding Group IA metals). These metals melt at temperatures between 650°C and 850°C (except for beryllium, m.p. 1280°C).

5 Calcium, strontium and barium form hydrides, $M^{2+}2H^-$.

6 All the metals, except for beryllium, form nitrides by direct combination. Reaction of the nitride with water produces ammonia. For example,

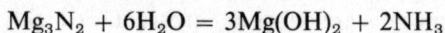

$$Mg_3N_2 + 6H_2O = 3Mg(OH)_2 + 2NH_3$$

7 The oxides and hydroxides of the metals are strongly basic. (Beryllium oxide is amphoteric.)

8 Because of the higher charge-to-volume ratio of the cations, the salts of these metals are more deliquescent, or more highly hydrated than the corresponding alkali metal salts. This feature decreases as the charge-to-volume ratio decreases with later members of the group; for example, barium chloride crystals contain two molecules of water of crystallization, whereas calcium chloride is deliquescent.

9 All the metals are very reactive, and decompose water (magnesium reacts with boiling water, the others react in the cold) to produce hydrogen.

10 Beryllium and magnesium form soluble sulphates; the remaining metals form insoluble sulphates.

11 Calcium, strontium and barium give flame colorations.

12 Beryllium and magnesium form many organo-metallic compounds.

13 Beryllium forms many basic salts.

Magnesium

Ore: magnesium carbonate, as magnesite, $MgCO_3$, and dolomite, $CaCO_3 . MgCO_3$.

$E^\ominus$ **value**: $-2 \cdot 37$ V.

Reaction of metal with:

(a) *Air* (on heating),

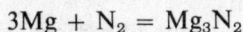

$$2Mg + O_2 = 2MgO$$

$$3Mg + N_2 = Mg_3N_2$$

(b) *Water* (at boiling point)

$$Mg + H_2O = MgO + H_2$$

(c) *Acids* (cold, dilute),

$$Mg + 2H^+ = Mg^{2+} + H_2$$

(d) *Halogeno-alkanes* (alkyl halides) (in dry ethoxy-ethane (ether), on gentle warming),

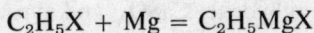

$$C_2H_5X + Mg = C_2H_5MgX$$

(Grignard reagents: X = Cl, Br, I).

Outline of the Main Reactions of Magnesium

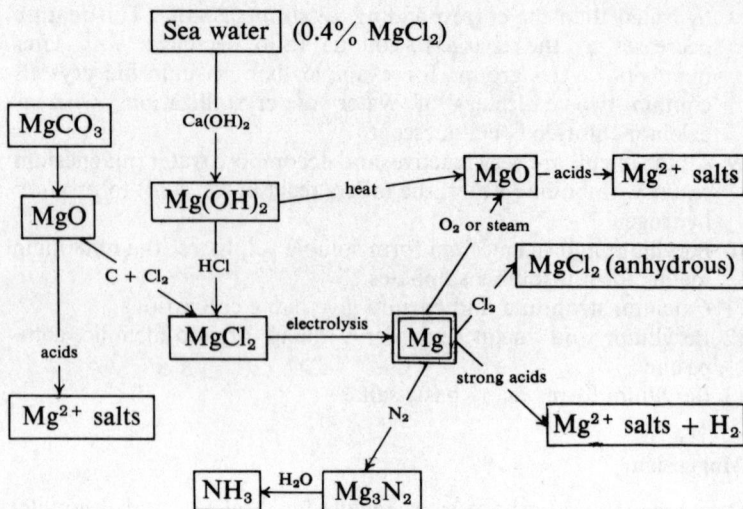

Similarities Between Magnesium and Lithium (Diagonal Relationships —Fajan's Rules)

Both metals:

1 form normal oxides only, on heating in oxygen,
2 form nitrides by direct union of the elements,
3 form deliquescent salts,
4 show thermal decomposition of carbonates and nitrates,
5 form sparingly soluble carbonates and phosphates,
6 form organo-metallic compounds,
7 produce salts which are soluble in organic solvents.

Calcium

Ores: carbonate—as calcite, Iceland spar, limestone, chalk and marble, $CaCO_3$
phosphate—as rock phosphate $Ca_3(PO_4)_2$
fluoride—as fluorspar, CaF_2
sulphate—as anhydrite, $CaSO_4$, and gypsum, $CaSO_4.2H_2O$

$E^\ominus$ **value:** $-2·87$ V

Reaction of the metal with:

(a) *Air* (on heating),

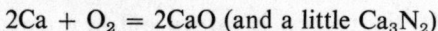

$$2Ca + O_2 = 2CaO \text{ (and a little } Ca_3N_2)$$

(b) *Water* (in the cold),

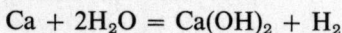

$$Ca + 2H_2O = Ca(OH)_2 + H_2$$

(c) *Hydrogen* (on heating and under pressure),

$$Ca + H_2 = CaH_2$$

Outline of the Main Reactions of Calcium

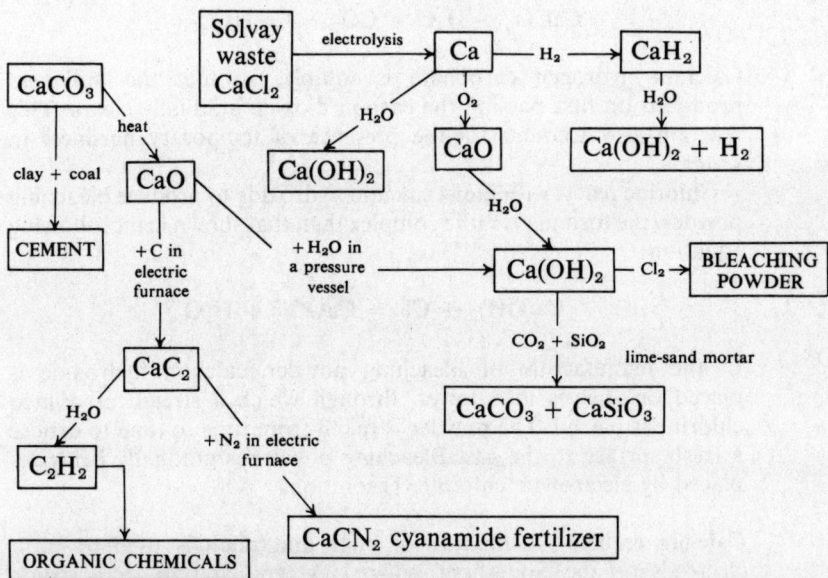

Calcium oxide This is made by heating calcium carbonate in a vertical gas or oil-fired furnace:

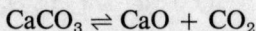

$$CaCO_3 \rightleftharpoons CaO + CO_2$$

Calcium oxide reacts with chlorine above 300°C:

$$2CaO + 2Cl_2 = 2CaCl_2 + O_2$$

101

Calcium oxide reacts readily with water to produce the hydroxide:

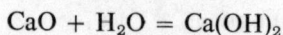

$$CaO + H_2O = Ca(OH)_2$$

Industrially, the process is carried out in a closed pressure vessel, and the product has a very fine particle size.

Calcium hydroxide (hydrated lime, or slaked lime) is used in the chemical industry, in paper making, tanning, as a fertilizer, and in lime-sand mortar.

Calcium hydroxide is sparingly soluble in water. The solution is called lime water, and is used in testing for carbon dioxide:

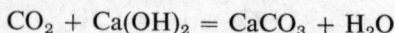

$$CO_2 + Ca(OH)_2 = CaCO_3 + H_2O$$

With excess carbon dioxide,

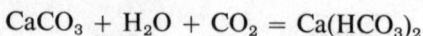

$$CaCO_3 + H_2O + CO_2 = Ca(HCO_3)_2$$

Calcium hydrogen carbonate is soluble, so that the milkiness produced on first passing the carbon dioxide gradually clears. This reaction also accounts for the presence of temporary hardness in water.

Chlorine reacts with moist calcium hydroxide to produce bleaching powder (the formula is more complex than that shown in the following equation)

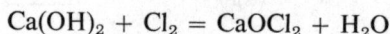

$$Ca(OH)_2 + Cl_2 = CaOCl_2 + H_2O$$

In the manufacture of bleaching powder, calcium hydroxide is placed on shelves in a tower, through which a stream of diluted chlorine is passed. The powder is raked from time to time to expose a fresh surface to the gas. Bleaching powder is gradually being replaced by electrolytic chlorate (I) solutions.

Calcium carbide A mixture of coke and pitch is used to form electrodes of the 'Soderberg' self-baking type, and an arc is struck between these electrodes and the graphite base of a cell containing calcium oxide. The reaction is

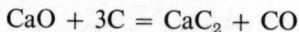

$$CaO + 3C = CaC_2 + CO$$

Calcium carbide is decomposed by water to form ethyne:

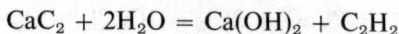

$$CaC_2 + 2H_2O = Ca(OH)_2 + C_2H_2$$

Ethyne can be used as a starting material for other organic chemicals. Calcium carbide also reacts with nitrogen when heated to over 1000°C:

$$CaC_2 + N_2 = CaCN_2 + C$$

The final mixture is called *cyanamide*, and is used as a fertilizer, and as a source of ammonia:

$$CaCN_2 + 3H_2O = CaCO_3 + 2NH_3$$

Cement This is a mixture of the silicates of calcium, iron and aluminium. The raw materials, consisting of cement rock, limestone, clay and gypsum, are fed into a rotary kiln fired by oil or powdered coal. The temperature rises to over 1300°C, and the final product is a clinker, which is finely ground before packaging.

8 The Chemistry of the d-Block Elements (The Transition Metals)

These elements make up a block of three rows in the Periodic Table as shown on page 7, Table 1, but only the first of these series, that running from scandium to zinc, will be discussed.

Since zinc does not use d-electrons in bonding, the chemistry of this element will be discussed separately.

General Properties of the Transition Metals

In addition to showing the normal metallic properties such as conduction of heat and electricity, forming electropositive ions, and having a high tensile strength, etc., these metals also show the following special characteristics:

1 Formation of coloured ions.
2 Magnetic properties.
3 Variable oxidation states.
4 Complex formation.

All these features can be related to the electronic structure of this series of elements.

The electronic configuration of the transition elements

Sc	(Ne) . $3d^1 4s^2$	
Ti	(Ne) . $3d^2 4s^2$	
V	(Ne) . $3d^3 4s^2$	
Cr	(Ne) . $3d^5 4s^1$	(out of sequence)
Mn	(Ne) . $3d^5 4s^2$	
Fe	(Ne) . $3d^6 4s^2$	
Co	(Ne) . $3d^7 4s^2$	
Ni	(Ne) . $3d^8 4s^2$	
Cu	(Ne) . $3d^{10} 4s^1$	(out of sequence)
Zn	(Ne) . $3d^{10} 4s^2$	

The Formation of Coloured Ions

It is possible for electrons to be promoted from one energy level to another. To do this, an amount of energy exactly equal to the difference in the energies between the two levels has to be given to the electron. For small energy differences, the amount of energy corresponds to that represented by a particular wavelength of visible light. When a substance absorbs a certain wavelength (for example, red) from white light, the transmitted or reflected light contains an 'unbalanced' proportion of another wavelength (blue, in this instance) and the substance appears coloured.

Just such a small gap between energy levels occurs with ions of the transition metals. It is thought that on the approach of negative ions (or the negative 'ends' of polar molecules) to a transition metal ion, a slight separation of the 3d-orbitals into two distinct energy levels ensues. This splitting of orbital energies is shown diagrammatically in Figure 50. A more complex type of splitting is apparent when the

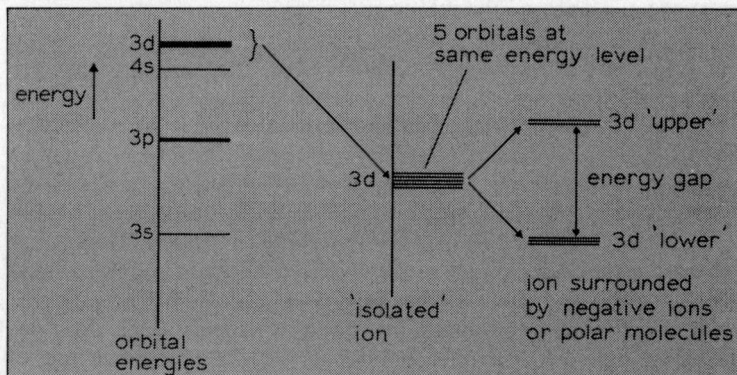

Fig. 50 Separation of 3d-orbital energies

distribution of negative ions (or polar molecules) round the transition metal ion is other than regularly octahedral or tetrahedral.

The separation of the 3d-orbital energies is small and depends on:

(a) the metal,
(b) the identity of the negative ions or polar molecules surrounding the transition metal ion.

Thus, the colour of $[Ni(H_2O)_6]^{2+}$ (green) is quite different from that of $[Ni(NH_3)_6]^{2+}$ (purple) [factor (b)], while the colour of nickel sulphate is quite different from that of copper sulphate [factor (a)].

It is clear that for an electronic transition to take place according to this scheme, there must be:

(a) at least one electron in the lower 3d-level,
(b) at least one vacancy in the upper 3d-level, so that the promoted electron can be accepted into the upper level.

Hence, Sc^{3+}, $3d^0$, $4s^0$, is colourless for the first reason, while both copper (I) and zinc (II) ions ($3d^{10}$), are colourless for the second reason.

Magnetic Properties

Most of the transition metal ions are *paramagnetic*, that is they attract magnetic lines of force. This effect is due to the presence of unpaired electron spins in the ion, since a rotating electron is equivalent to an electrical current and thus produces a magnetic effect. The magnetic moment μ of a transition metal ion is related to the number n of unpaired electrons by the relation:

$$\mu = \sqrt{\{n(n + 2)\}}$$

The units of magnetic moment are expressed in Bohr magnetons (B.M.).

Hence, a copper (I) ion, having a $3d^{10}$ structure, contains no unpaired electrons and has a magnetic moment of zero; this is confirmed by experiment. On the other hand, a copper (II) ion, with the configuration $3d^9$, has a single unpaired electron, and this should give rise to a magnetic moment of $\sqrt{3}$ or 1·73 B.M. The observed value is 1·9 B.M. *A diamagnetic* substance is one which does not have any unpaired electrons; such substances repel magnetic lines of force.

Variable Oxidation States

The factors governing the number of electrons lost during cation formation were outlined on p. 37. From this, it is evident that the ionization energy of the atom is an important factor in limiting the number of electrons eventually lost during ion formation. In general, the extra lattice energy gained as a result of the formation of an ion carrying a higher charge has to balance against the extra ionization energy required to remove the electrons necessary to give this charge. For the non-transition metals, a significant jump in ionization energy is noticed, after the removal of one electron in the case of a Group IA metal, or two electrons in the case of a Group IIA metal, and so on. With transition metals, such large jumps in ionization energies are not observed, since the energies of the 4s- and 3d-orbitals are not

106

widely different. Therefore, the extra lattice energy released when a cation carrying a higher charge is formed, is very finely balanced against the increase in ionization energy needed to produce that cation. Reference to the minor irregularities shown in the ionization energy chart (Figure 51) can indicate the preferred oxidation state.

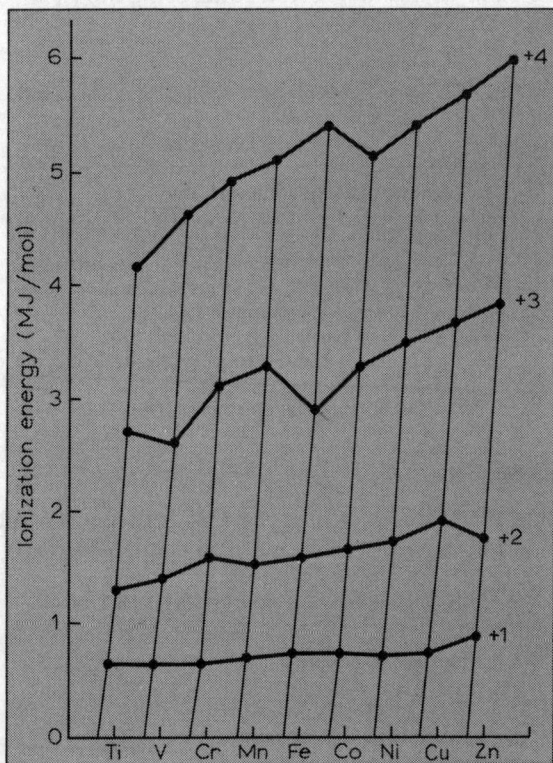

Fig. 51 Successive ionization energies of the transition metals

In general, the gap between the $+2$ and $+3$ ionization energies increases from left to right, showing that lower oxidation states would be preferred by the transition elements appearing at the end of the series. Small dips in the graph corresponding to the $+2$ oxidation state for zinc and manganese, and the $+3$ oxidation state for vanadium agree with the fact that these are the common oxidation states found with these elements. All the transition metals are capable of showing an oxidation state of $+2$, the gap between the $+1$ and the $+2$ state being greatest with copper, which corresponds to the fact that copper (I) salts are well known. Oxidation states in excess of $+3$ are thought to be associated with covalent rather than ionic bonding.

Using the relevant redox potentials (see p. 72), the relative stability of the various oxidation states may be assessed; the results are indicated in Table 13, the more stable states being shown in bold type.

TABLE 13 *Typical oxidation states of the transition metals*

Element	Oxidation state
Sc	**+3** (shows one oxidation state only)
Ti	$+2$ **+3** $+4$ ($+4$ state is covalent)
V	$+2$ **+3** **+4** **+5**
Cr	**+2** **+3** $+6$ ($+6$ in chromates and dichromates)
Mn	**+2** $+3$ $+4$ $+7$ ($+7$ in permanganate)
Fe	**+2** $+3$ $+6$ ($+6$ in ferrates)
Co	**+2** $+3$ ($+3$ particularly in complexes)
Ni	**+2** ($+2$, also $+1$ and zero found in nickel complexes)
Cu	$+1$ **+2**
Zn	**+2** (shows one oxidation state only)

Complex Formation

Investigation of the product of a reaction between two substances which would not normally be expected to combine further; such as

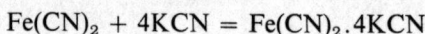

$$Fe(CN)_2 + 4KCN = Fe(CN)_2 . 4KCN$$

shows that:

(a) no free iron (II) ions are present,
(b) no free cyanide ions are present,
(c) free potassium ions are present,
(d) a total of five ions are present.

These observations are explained by *complex formation*; that is the combination of the iron (II) ions and the cyanide ions to form a *complex nucleus*:

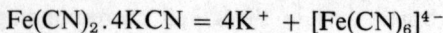

$$Fe(CN)_2 . 4KCN = 4K^+ + [Fe(CN)_6]^{4-}$$

A complex nucleus (distinguished by writing the formula inside square brackets) consists of a central ion (or atom) combined with several groups known as *ligands*. The number of ligands combined with the central ion is called the *complex coordination number*.

The molecules or ions which commonly function as ligands are:

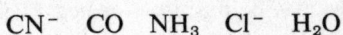

$$CN^- \quad CO \quad NH_3 \quad Cl^- \quad H_2O$$

A common feature of the electronic structure of all these ligands, is the possession of a *lone pair* of electrons, that is two spin-paired valency electrons not involved in bonding (Figure 52). These lone

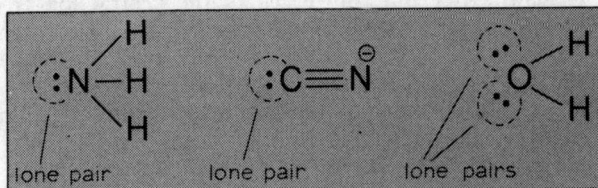

Fig. 52 Structures of some ligands

pairs of electrons take part in forming a coordinate link, that is, a covalent bond in which both the shared electrons are provided by one atom only. For example,

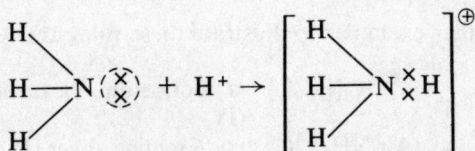

Coordinate bonds form when:

1 a ligand possesses at least one lone pair of electrons,
2 the central atom or ion is capable of

(a) attracting the lone pair, and,
(b) accommodating the lone pair in its own vacant valence level orbitals.

Transition metal ions satisfy these requirements, since they are small and highly charged. They have empty 4s- and 4p-orbitals in addition to the vacancies within the 3d-level, as this level is usually only partly filled.

Complexes are sometimes referred to as *coordination compounds*, in view of the nature of the bonds from ligand to metal ion.

109

Rules for Naming Complexes

The complex nucleus may be positively or negatively charged, and in some cases it may not carry any charge at all. In all cases the name of the cation, whether simple or complex, is given first.

Simple cations, such as K^+ are given their usual names.

For a complex cation, the names of the ligands are given first in alphabetical order. A suitable prefix is used to show the number of ligands present. After this, the name of the central ion is given, followed by its oxidation state in brackets. For example,

$[Cu(NH_3)_4]SO_4$	is tetrammine copper (II) sulphate
$[Cr(NH_3)_5Cl]Cl_2$	is pentamminechlorochromium (III) chloride.

In the case of complex anions, the method of nomenclature is the same as for complex cations, with the further modification that the name of the central ion is altered to end in -ate. For example,

$$K_4[Fe(CN)_6] \quad \text{is potassium hexacyanoferrate (II).}$$

Negatively charged ligands take the ending -o; for example, cyanide ion ligands are named cyano- in complexes. Neutral ligands have no special ending; for example, NH_3 is an ammine ligand, and H_2O is aquo.

Some further examples to illustrate these rules are:

$K_2[PtCl_6]$	is potassium hexachloroplatinate (IV)
$[Ag(NH_3)_2]Cl$	is diammine silver (I) chloride.
$[Co(NH_3)_3(NO_2)(CN)(Cl)]$	is triamminechlorocyanonitrito-cobalt (III)

Multi-coordinating Ligands

Some ligands are capable of supplying more than one lone pair for coordination. Ethane-1,2-diamine is an example of this, the two lone pairs being carried by the two nitrogen atoms. In the formula for the complex, ethane-1,2-diamine is usually denoted by the abbreviation 'en'.

Ligands which can provide two lone pairs for coordination are said to be *bidentate* ligands (i.e., holding pincer-like between two teeth). When they are bonded to a central ion through the formation of two coordinate bonds, a ring structure is generally produced. This is called chelation and the complex ion is known as a chelate. Chelate complexes are more stable than the corresponding complexes formed

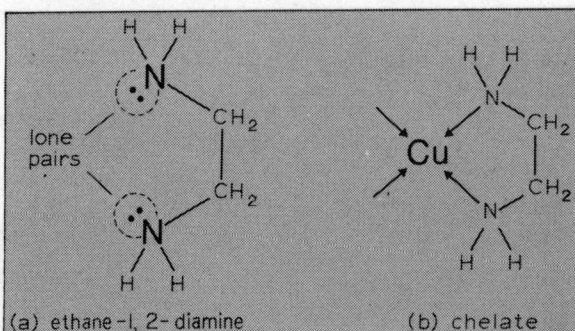

Fig. 53 Chelation

by monodentate ligands (i.e., those forming a single coordinate link).

Further examples of bidentate ligands are the ethanedioate ion and the glycinate ion (derived from glycine—amino ethanoic acid), the structures of which are shown in Figure 54.

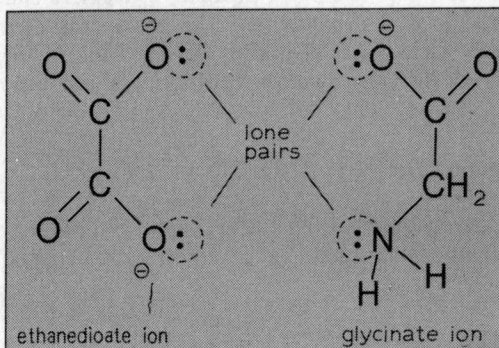

Fig. 54 Bidentate ligands

A specially important ligand is the ion derived from ethane-1,2-diamine tetra-ethanoic acid (abbreviated to EDTA). This ligand is capable of supplying up to six lone pairs to the central metal ion. Hence, EDTA is able to form very stable complexes with the transition metal ions and some non-transition metal ions. In addition it is found that 1 mole of metal ion complexes with 1 mole of EDTA.

There is a degree of flexibility within this molecule which allows the six lone pairs to approach the central metal atom octahedrally (see p. 115).

111

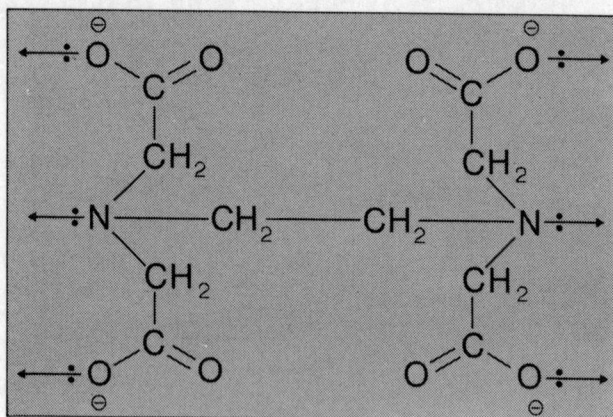

Fig. 55 EDTA

The Charge Carried by Complex Ions

The complex nucleus may carry a positive or negative charge, and it is then referred to as a *complex ion*. The net overall charge on the complex ion is arrived at by taking the algebraic sum (i.e., with regard to sign) of the charges of the central metal ion and the several ligands.

Neutral ligands are: NH_3, H_2O, CO and en.

Singly negatively charged ligands are: Cl^-, Br^-, CN^- and OH^-. Doubly negatively charged ligands are: ethanedioate, $C_2O_4^{2-}$ and sulphate SO_4^{2-}.

Some examples:

In the complex ion

$$[Co^{III}(NH_3)_5Cl]$$

the charge of $+3$ on the cobalt is reduced by 1 due to the singly negatively charged chloride ion. The ammonia ligands are neutral, which produces a net charge of $+2$ on the complex ion. Hence $[Co^{III}(NH_3)_5Cl]^{2+}$ is a complex cation which gives rise to such salts as

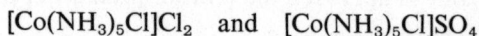

$$[Co(NH_3)_5Cl]Cl_2 \quad \text{and} \quad [Co(NH_3)_5Cl]SO_4$$

By similar reasoning, the net charge on the complex ion

$$[Fe^{III}(CN)_6]$$

is

$$+3 - 6 = -3$$

Hence $[Fe(CN)_6]^{3-}$ is a complex anion which gives rise to such salts as $K_3[Fe(CN)_6]$.

Complex Coordination Number and Multidentate Ligands

The coordination number of copper is four in both of the following complexes:

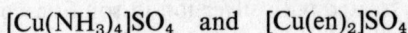

$$[Cu(NH_3)_4]SO_4 \quad \text{and} \quad [Cu(en)_2]SO_4$$

In each case, four coordinate bonds are formed, although the number of ligands attached to the central metal ion differs. To allow for the cases where multidentate ligands are involved, the coordination number of a complex is better expressed as the number of coordinate bonds formed between the central metal ion and the ligands.

Isomerism in Complexes

Since the ligands are attached to the central metal ion by coordinate covalent bonds, the possibility of *isomerism* (i.e., the existence of two or more different structures corresponding to the same molecular formula) arises. Isomerism occurs widely among organic compounds, but some simple cases of isomerism in complexes are outlined as follows.

Ionization isomerism The charge carried by the complex ion

$$[Co^{III}(NH_3)_5(SO_4)]$$

is $+3 - 2 = +1$. A typical salt formed by this ion is the bromide, the formula for the salt being

$$[Co(NH_3)_5(SO_4)]Br$$

If another complex ion can be produced of formula

$$[Co^{III}(NH_3)_5Br]$$

113

the charge on this ion is $+2$. A typical salt might be

$$[Co(NH_3)_5Br]SO_4$$

Both these salts have the same molecular formula, but they are structurally quite different, and are thus isomers.

Hydrate isomerism This is rather similar to ionization isomerism described above. It results from water molecules either behaving as ligands or functioning as molecules of water of crystallization.

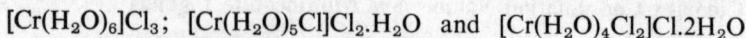

$[Cr(H_2O)_6]Cl_3$; $[Cr(H_2O)_5Cl]Cl_2.H_2O$ and $[Cr(H_2O)_4Cl_2]Cl.2H_2O$

are isomeric. They can be distinguished by the fact that one mole of complex precipitates three, two or one mole of silver chloride respectively when treated with silver ion in solution corresponding to the number of chloride ions found as anions outside the complex nucleus.

Coordination isomerism This occurs when both the cation and anion are complex, for example, $[Co(NH_3)_6][Cr(CN)_6]$ and $[Cr(NH_3)_6]$ $[Co(CN)_6]$. The two forms can be distinguished by passing a solution of the complex through an ion exchange column, in which the cation is exchanged for H^+. The eluant is then examined to identify which metal and which ligand is present in the complex anion which passes unchanged through the ion exchange column.

Stereoisomerism This may take two forms, geometrical isomerism and optical isomerism (See Part 2.)

Before giving examples of this type of isomerism, we have to consider the hybridization of the orbitals and the structure of the complex nucleus resulting from this hybridization. Taking the case of a chromium complex, Cr has the electronic configuration $1s^2 2s^2 2p^6 3s^2 3p^6 3d^5 4s^1$. After ionization to become Cr^{3+} the configuration of the ion is $1s^2 2s^2 2p^6 3s^2 3p^6 3d^3$. This leaves two 3d orbitals empty, as are the 4s and the 4p.

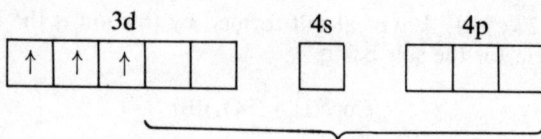

These six orbitals hybridize to form six equivalent d^2sp^3 hybrid orbitals.

Each hybrid orbital can accept a lone pair of electrons, so that six monodentate ligands can bond to the central metal atom. For example

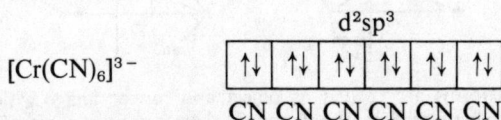

d^2sp^3

$[Cr(CN)_6]^{3-}$

↑↓	↑↓	↑↓	↑↓	↑↓	↑↓

CN CN CN CN CN CN

The six hybrid orbitals are distributed evenly in space—that is octahedrally, so that the shape of the complex nucleus is, in this case, octahedral.

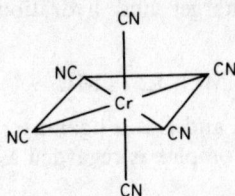

Alternative hybridizations result in tetrahedral (sp^3 hybridization) or square planar (dsp^2) structures for the complex nucleus. The fact that the complex nucleus has a definite shape allows for the possibility of geometrical and optical isomerism.

Geometrical isomerism This type is found only in square planar or octahedral complexes. For example, *cis* and *trans* isomers of the compound $[Co(en)_2(NO_2)_2]^+Cl^-$ have been isolated.

(en = ethane−1,2−diamine)

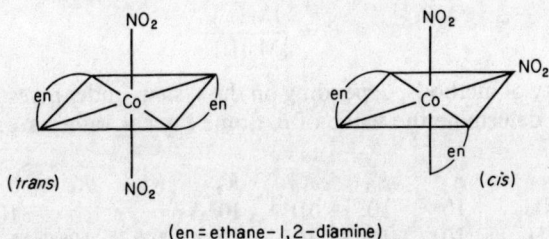

The *trans* form has no dipole moment

Optical isomerism The *cis* form of the compound mentioned above is optically active and should therefore be capable of resolution.

115

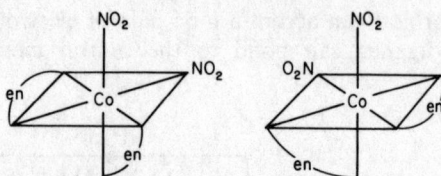

These two forms are related as object and mirror image which are not superimposable (see Part 2) and should therefore be capable of resolution

Stability of Complexes

The stability of a complex is a measure of its resistance to the replacement of one ligand by another, although the term is most often used to show the tendency of a metal ion in aqueous solution to form a complex. (Ignoring charges and hydration effects, this can be expressed by:

$$M + L_n \rightleftharpoons ML_n$$

where M is the metal ion and L is a ligand.)

The formation of the complex is regarded as a series of steps:

(a) $M + L \rightleftharpoons ML$
(b) $ML + L \rightleftharpoons ML_2$
    ⋮
(c) $ML_{n-1} + L \rightleftharpoons ML_n$.

For each step, an equilibrium constant (K) can be evaluated

$$K_1 = \frac{[ML]}{[M][L]} \qquad K_2 = \frac{[ML_2]}{[ML][L]} \text{ etc.}$$

and the product of the successive stability constants (K_1, K_2 etc.) is called the **overall stability constant** β.

Therefore,
$$\beta = \frac{[ML_n]}{[M][L]^n}$$

A variety of methods, depending on the system under investigation, is used to determine the value of β. Some typical values are:

	K_1	K_2	K_3	K_4	K_5	K_6	β
Cu^{2+}/NH_3	$10^{4.3}$	$10^{3.5}$	$10^{3.1}$	$10^{2.2}$			$10^{13.1}$
Ni^{2+}/NH_3	10^3	$10^{2.2}$	$10^{1.7}$	$10^{1.2}$	$10^{0.75}$	$10^{0.05}$	$10^{8.9}$

Because of the high values of β, $p\beta$ ($= \log_{10}\beta$) values are quoted. For $[Cu(NH_3)_4]^{2+}$ $p\beta = 13.1$; for $[Cd(NH_3)_4]^{2+}$ $p\beta = 7.0$; and $p\beta$ for $[Fe(CN)_6]^{3-}$ is 30.

Outline of the Reactions of Selected Transition Metals

Chromium

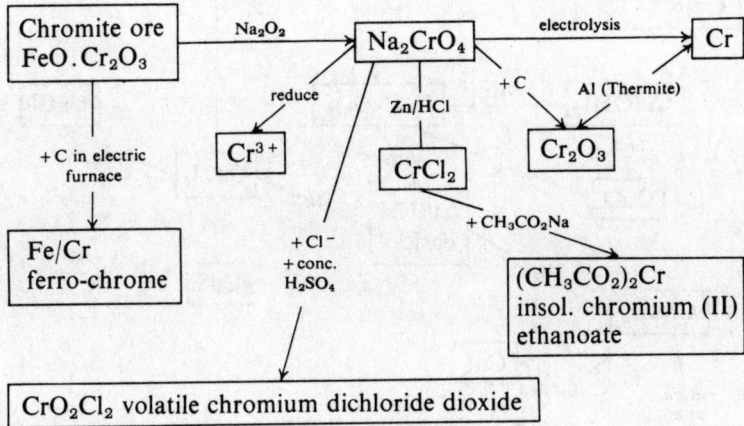

Manganese

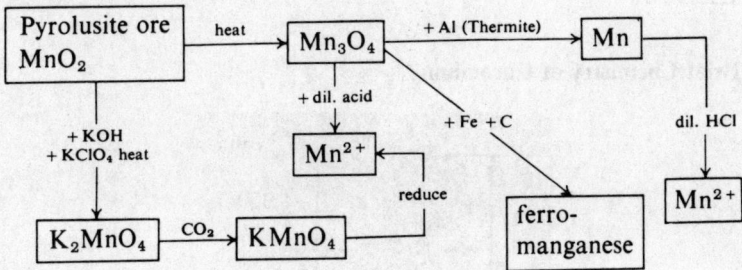

Iron

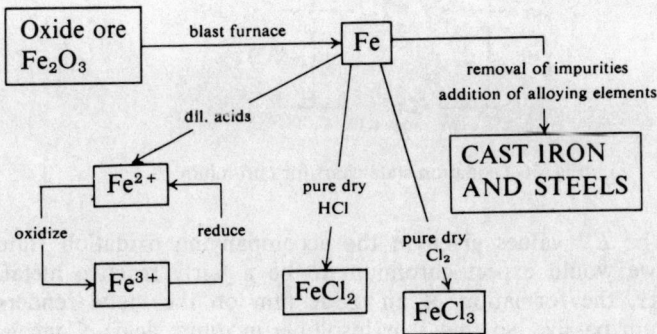

Copper

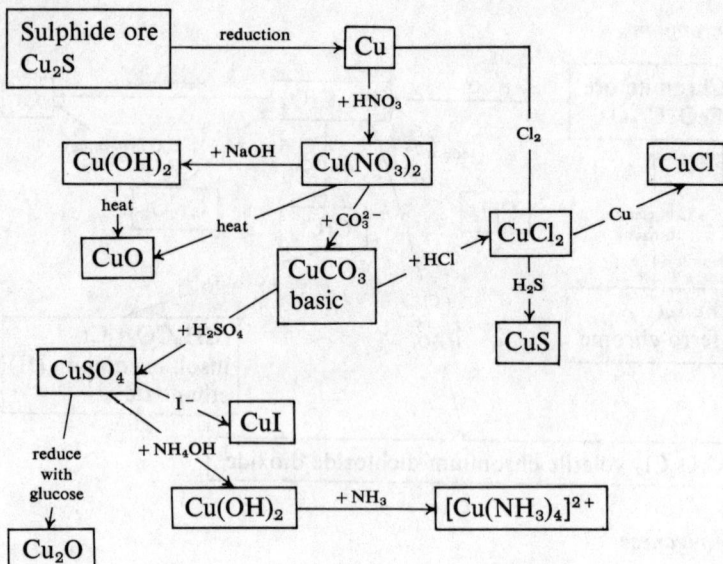

Brief Chemistry of Chromium

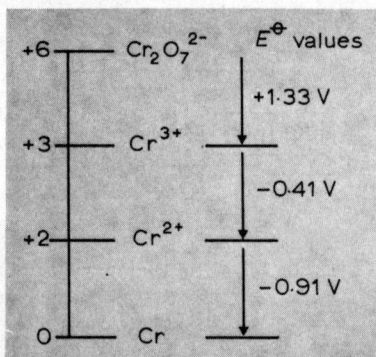

Fig. 56 Oxidation state chart for chromium

From the $E^{\ominus}$ values given in the accompanying oxidation state chart, we would expect chromium to be a fairly reactive metal. However, the formation of an oxide film on the metal renders chromium passive, so that it is insoluble in dilute acids. Concentrated nitric acid renders chromium passive, but the metal dissolves in hot concentrated hydrochloric or sulphuric acids:

$$2HCl + Cr = CrCl_2 + H_2$$

In the presence of atmospheric oxygen, oxidation of the product takes place:

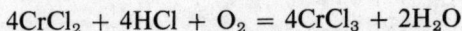

$$4CrCl_2 + 4HCl + O_2 = 4CrCl_3 + 2H_2O$$

From the oxidation state chart, chromium in oxidation state $+6$ (as chromate or dichromate) is powerfully oxidizing. Sodium dichromate (VI) is prepared directly from chromite ore by roasting with sodium carbonate in air:

$$8Na_2CO_3 + 4FeCr_2O_4 + 7O_2 = 8Na_2CrO_4 + 2Fe_2O_3 + 8CO_2$$

Sodium chromate is formed and this is dissolved out of the reaction mixture with water. Acidification with concentrated sulphuric acid produces a solution containing sodium dichromate (VI), which crystallizes out on evaporating and cooling:

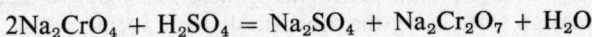

$$2Na_2CrO_4 + H_2SO_4 = Na_2SO_4 + Na_2Cr_2O_7 + H_2O$$

Ammonium dichromate (VI) decomposes spectacularly on heating:

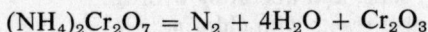

$$(NH_4)_2Cr_2O_7 = N_2 + 4H_2O + Cr_2O_3$$

Chromium dichloride dioxide (chromyl chloride) is produced by heating a chloride with a dichromate and concentrated sulphuric acid:

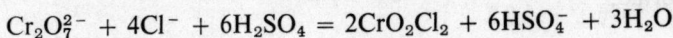

$$Cr_2O_7^{2-} + 4Cl^- + 6H_2SO_4 = 2CrO_2Cl_2 + 6HSO_4^- + 3H_2O$$

Chromium dichloride dioxide is a dark-red liquid, boiling point 117°C. Its production is used as a test for chlorides (in the presence of bromides and iodides, which do not form similar compounds) since the vapours hydrolyse readily when bubbled through water:

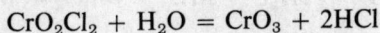

$$CrO_2Cl_2 + H_2O = CrO_3 + 2HCl$$

This solution gives a positive test for chromate. On mixing a solution of chromium trioxide or acidified dichromate with hydrogen peroxide, a blue solution containing 'perchromic' acid (possibly

119

CrO_5) is produced. This compound is unstable in aqueous acid, but is more stable in ethoxy-ethane (ether), into which it is often extracted in performing this test.

Brief Chemistry of Manganese

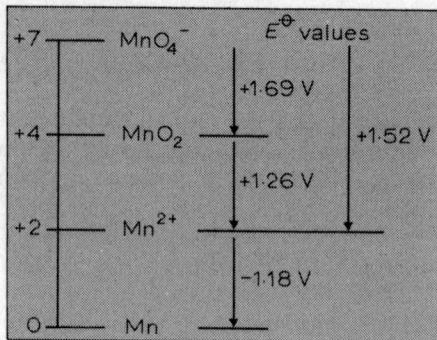

Fig. 57 Oxidation state chart for manganese

From the oxidation state chart, manganese is a reactive metal, the +2 oxidation state being the most stable ionic form.

Manganese liberates hydrogen from steam, at red heat, and reacts readily with dilute hydrochloric and sulphuric acids, forming manganese (II) salts, and liberating hydrogen.

Manganese also reacts with chlorine:

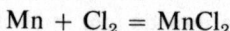

$$Mn + Cl_2 = MnCl_2$$

while a similar reaction with fluorine produces manganese (III) fluoride:

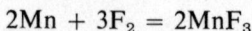

$$2Mn + 3F_2 = 2MnF_3$$

This is the only trihalide of manganese, and it reacts with water to form manganese dioxide and manganese (II) fluoride:

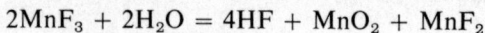

$$2MnF_3 + 2H_2O = 4HF + MnO_2 + MnF_2$$

Manganese dioxide is an oxidizing agent, as seen from the oxidation state chart. In dilute sulphuric acid, it oxidizes iron (II) sulphate:

$$2FeSO_4 + 2H_2SO_4 + MnO_2 = MnSO_4 + Fe_2(SO_4)_3 + 2H_2O$$

120

Manganese dioxide is acidic, and on fusing with sodium or potassium hydroxide, in the presence of a strong oxidizing agent, or with free access to air, manganate (VI) is produced:

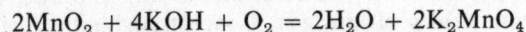

$$2MnO_2 + 4KOH + O_2 = 2H_2O + 2K_2MnO_4$$

The dark-green product, potassium manganate (VI), is stable in excess alkali. On dilution, or acidification, or on passing carbon dioxide into the solution, manganese dioxide is deposited and potassium manganate (VII) is produced:

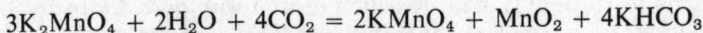

$$3K_2MnO_4 + 2H_2O + 4CO_2 = 2KMnO_4 + MnO_2 + 4KHCO_3$$

Potassium manganate (VI) can be oxidized to manganate (VII) without deposition of manganese dioxide by using chlorine, or electrolytically.

Manganate (VII) ions are strong oxidizing agents, although their oxidizing power depends on the pH of the solution. In strong acid solution:

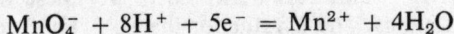

$$MnO_4^- + 8H^+ + 5e^- = Mn^{2+} + 4H_2O$$

In slightly alkaline solution:

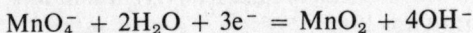

$$MnO_4^- + 2H_2O + 3e^- = MnO_2 + 4OH^-$$

Brief Chemistry of Iron

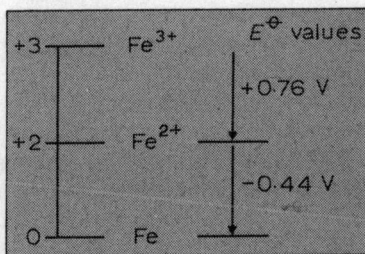

Fig. 58 Oxidation state chart for iron

From the oxidation state chart for iron, the metal is moderately reactive, decomposing steam at red heat:

$$3Fe + 4H_2O \rightleftharpoons Fe_3O_4 + 4H_2$$

121

Hydrochloric and dilute sulphuric acids react readily with iron, liberating hydrogen. Hot, concentrated sulphuric acid reacts to produce sulphur dioxide and iron (II) sulphate. Dilute nitric acid forms iron (II) nitrate in the cold, and iron (III) nitrate when hot, together with oxides of nitrogen. Concentrated nitric acid renders iron passive.

On heating, iron reacts with chlorine to form iron (III) chloride:

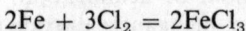

$$2Fe + 3Cl_2 = 2FeCl_3$$

To produce anhydrous iron (III) chloride, the apparatus shown in Figure 61 (p. 133) is used. Reaction of iron with pure dry hydrogen chloride in a similar apparatus produces anhydrous iron (II) chloride:

$$Fe + 2HCl = FeCl_2 + H_2$$

To oxidize iron (II) to iron (III) oxidizing agents having an $E^\ominus$ value greater than $+0.76V$ are required. From Table 12 (p. 72), these include silver (I) ions, bromine, chlorine, dichromate (VI) and manganate (VII). To reduce iron (III) to iron (II) the reducing agent must have an $E^\ominus$ value less than $+0.76$ V. These include iodide ion, and tin (II) ions. The E value for the Fe^{3+}/Fe^{2+} couple is again pH-dependent, and falls from the standard value $E^\ominus$ of $+0.76$ V at pH zero, to a negative value in slightly alkaline solution. This explains why hydrogen peroxide can oxidize iron (II) in solutions in which the acid concentration is low.

Iron and Steel Manufacture

The product of the blast furnace is *pig iron*, and this contains 3–4 per cent carbon and other impurities. Remelting pig iron improves the grain structure, so that the resulting metal can be used for making castings. This is *cast iron*, and it contains 2·5–3·5 per cent of carbon in the form of graphitic flakes deposited in between the iron crystals. Cast iron is cheap, has a low melting point and has a high strength under compression. However, it is brittle, and cannot be forged or hardened.

Remelting pig iron in a reverberatory furnace in the presence of iron oxide, gradually oxidizes the impurities. As the iron increases in purity, the melting point rises, and at 1550°C the product contains less than 0·1 per cent of carbon. This is *wrought iron*. It is soft, bends easily and can withstand the impact of a sudden extensive strain. Hence it is used for decorative work and for hooks, couplings and chains. It also resists corrosion better than any other form of iron.

Steel is an alloy of iron and carbon—that is the carbon is held in solution within the iron crystals, and no free graphite is found along the grain boundaries. The amount of carbon present, which seldom exceeds 1·7 per cent, determines the grade of plain steel:

(a) Mild or low carbon steel contains 0·1–0·3 per cent carbon. This is a ductile material used for making tubes, rivets, structural work, forgings, pressings, etc.
(b) Medium steel (0·3–0·7 per cent carbon) is a material that can be hardened and tempered, and is used for making axles, driving shafts, springs and ropes, etc.
(c) High carbon steel (0·7–1·5 per cent carbon) can be hardened, and is used in making tools, chisels, files and razor blades, etc.

In addition to the plain steels, alloying elements are added (in the form of a master iron alloy such as ferro-chrome or ferro-vanadium) to give the steel special qualities. The addition of chromium improves the tensile strength of steel as well as its corrosion resistance. Nickel is useful in a similar way. Tungsten, copper, cobalt, vanadium and manganese are frequently used in steel alloys.

The conversion of iron into steel involves the following steps:

(a) removal of the impurities by oxidation,
(b) addition of alloying elements.

Modern methods of steelmaking use oxygen for the rapid removal of impurities in the iron. (An air blast contains nitrogen, which causes brittleness in steel.) One of these processes is the *Linz and Donawitz* (L.D.) process, in which a high-pressure jet of oxygen is

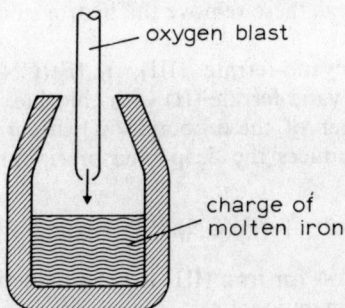

Fig. 59 L.D. converter

held about a metre above the surface of the molten iron in a converter. In this way, the impurities are removed in 20 to 30 minutes. Modifications of this process (such as the *Kaldo* and *Rotor* processes) use a rotating cylinder, of up to 100 tonnes capacity, inclined at an angle to the horizontal, and two oxygen jets are employed. Rotation of the cylinder agitates the charge, and this assists in the removal of the impurities. The older *open hearth* furnaces produce plain steel by melting a charge of pig iron, scrap iron and iron oxide in a shallow well of a furnace heated by burning a mixture of gas and air. The iron oxide slowly oxidizes the impurities, and an oxygen lance is used for the final removal of the carbon. The *Ajax furnace* is a modern type of open hearth furnace in which a jet of very high-pressure oxygen is blown into the charge. The heat released by the oxidation is sufficient to keep the charge molten.

Electric arc and induction furnaces are also used in steelmaking, and an oxygen lance is again employed for the final removal of impurities. Lime is often added to help in the removal of impurities in the form of a slag.

Alloy steels are made by adding the alloying elements (in the form of a master alloy) either as the final stage after the removal of the impurities by one of the above methods, or, more usually, to a remelted batch of plain steel heated in a small electric furnace.

Complexes of Iron

Potassium hexacyano-ferrate (II), $K_4[Fe(CN)_6]$, is one of the commonest complex compounds of iron. It is prepared by adding excess cyanide ion to a solution of an iron (II) salt:

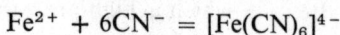

$$Fe^{2+} + 6CN^- = [Fe(CN)_6]^{4-}$$

It is manufactured from the iron oxide 'purifiers' used in the manufacture of coal gas, as these remove the hydrogen cyanide present in the crude coal gas.

Potassium hexacyano-ferrate (III), $K_3[Fe(CN)_6]$, is made by oxidizing the hexacyano-ferrate (II) with chlorine.

Reaction of either of these complexes with iron in the 'other' oxidation state produces the deep blue precipitate, Prussian blue. For example:

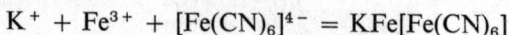

$$K^+ + Fe^{3+} + [Fe(CN)_6]^{4-} = KFe[Fe(CN)_6]$$

Another sensitive test for iron (III) is the production of a deep-red colour with thiocyanate ion:

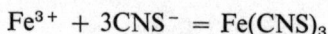

$$Fe^{3+} + 3CNS^- = Fe(CNS)_3$$

124

Brief Chemistry of Copper

From the oxidation state chart, copper is an unreactive metal, and is stable to air and water; also it does not liberate hydrogen from acids.

Dilute nitric acid attacks copper:

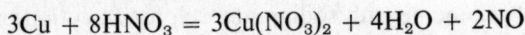

$$3Cu + 8HNO_3 = 3Cu(NO_3)_2 + 4H_2O + 2NO$$

and with concentrated acid, the reaction is vigorous:

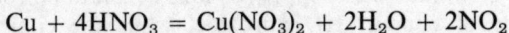

$$Cu + 4HNO_3 = Cu(NO_3)_2 + 2H_2O + 2NO_2$$

Concentrated sulphuric acid, when hot, attacks copper with the evolution of sulphur dioxide and the production of copper (I) and copper (II) sulphides.

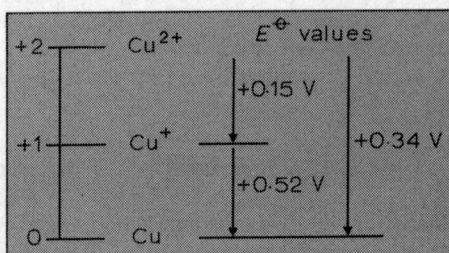

Fig. 60. Oxidation state chart for copper

Reference to the oxidation state chart shows that copper (I) is an oxidizing agent capable of oxidizing itself to copper (II). Thus, under standard conditions, copper (I) in aqueous solution undergoes disproportionation:

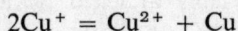

$$2Cu^+ = Cu^{2+} + Cu$$

Only when copper (I) is present in a complex, or as an insoluble salt is this ion stable in aqueous acid solutions.

Copper reacts with chlorine (heat is necessary if the metal is not finely divided) directly:

$$Cu + Cl_2 = CuCl_2$$

Copper dissolves in a solution of copper (II) chloride, and a precipitate of copper (I) chloride results when the reaction mixture is poured into water:

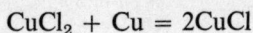

$$CuCl_2 + Cu = 2CuCl$$

125

Addition of iodide ion to a copper (II) salt in solution gives a precipitate of copper (I) iodide, and iodine is liberated:

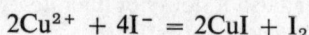

$$2Cu^{2+} + 4I^- = 2CuI + I_2$$

Titration of the iodine with sodium thiosulphate solution provides a method of determining the concentration of copper (II) ions in solution.

Addition of ammonia to copper (II) ions in solution produces first a precipitate of copper (II) hydroxide, while the complex tetra-ammine copper (II) ion is formed with excess ammonia:

$$Cu^{2+} + 2NH_4OH = Cu(OH)_2 + 2NH_4^+$$

Then,

$$Cu(OH)_2 + 4NH_3 = [Cu(NH_3)_4]^{2+} + 2OH^-$$

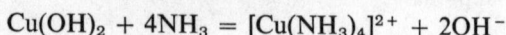

The best-known copper salt is copper (II) sulphate, which may be prepared by the action of sulphuric acid on the hydroxide, oxide or basic carbonate:

$$CuO + H_2SO_4 = CuSO_4 + H_2O$$

The salt crystallizes out with five molecules of water of crystallization, of which four are lost on heating to 100°C. The last molecule of water is removed only on strong heating (to 250°C).

Comparison of Copper, Silver and Gold

1 All three are 'noble' metals.
2 Attacked by nitric acid and hot concentrated sulphuric acid. Gold is only appreciably attacked by aqua regia (nitric and hydrochloric acids in the ratio 1:3).
3 Normal oxidation states are: copper $+2$, silver $+1$, and gold $+3$.
4 The oxides of these metals are easily reduced.
5 Readily form complexes, $[Ag(NH_3)_2]^+$ $[Au(CN)_2]^-$.
6 Although copper and silver appear to form some simple ionic compounds, gold compounds are probably covalent.

9 Zinc and Mercury

GROUP IIB: ZINC, CADMIUM AND MERCURY

General Properties

1 These metals immediately follow a transition series. However, the formation of complexes is the only typical transition metal property shown.

2 They have a weak metallic bonding system, and hence lower melting and boiling points than the corresponding transition elements. Mercury is a liquid at room temperature.

3 The oxidation state of all the metals is $+2$. In addition, mercury again shows a reluctance to use the outer electrons in bonding, and can exhibit an oxidation state of $+1$. In such compounds, two mercury atoms are joined by a single covalent bond, to form the ion $(Hg-Hg)^{2+}$.

4 Zinc and cadmium both stand above hydrogen in the electrochemical series, and hence they both displace hydrogen from acids. Mercury is a noble metal.

5 Zinc oxide and hydroxide are amphoteric, and zinc itself reacts with caustic alkalies. Cadmium and mercury (II) oxides are both basic.

6 Although zinc fluoride is ionic, the other halides have low melting points, and are appreciably soluble in organic solvents. This suggests that the latter are covalent to a significant extent. Cadmium chloride and bromide melt above 500°C (this temperature is usually regarded as the dividing line between the melting points of ionic and covalent compounds), and they are not very soluble in organic solvents. This would suggest that these compounds are more ionic in character. Only mercury fluoride, nitrate and chlorate (VII) are regarded as ionic.

7 The degree of hydration of the compounds of these metals decreases from zinc to mercury. Zinc salts are highly hydrated, while mercury compounds are usually anhydrous.

8 All the metals are capable of forming organo-metallic compounds, mercury forming the widest range of these.

Outline of the Reactions of Zinc

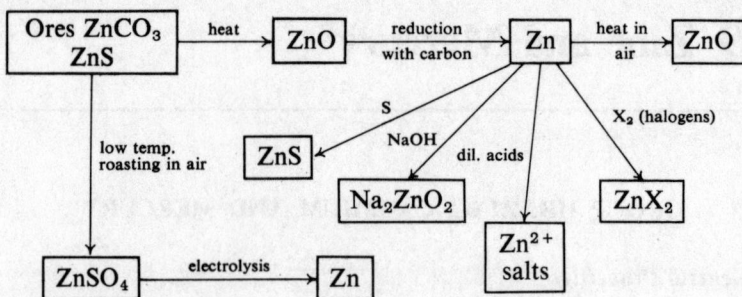

Brief Chemistry of Zinc

Ore: zinc sulphide as zinc blende (ZnS) and zinc carbonate as calamine ($ZnCO_3$).

$E^\ominus$ **value:** -0.76 V.

Reaction of the metal with:

(a) *Air:* Zinc burns when heated in air:

$$2Zn + O_2 = 2ZnO$$

(b) *Water:* No reaction except with steam at red heat, when hydrogen is liberated.

(c) *Acids:* Pure zinc is slow to react; impure zinc reacts readily with dilute hydrochloric and sulphuric acids:

$$Zn + 2H^+ = Zn^{2+} + H_2$$

Nitric acid (concentrated and dilute) attacks zinc to produce zinc nitrate and oxides of nitrogen. With very dilute nitric acid, a range of products is formed, including ammonium nitrate. Concentrated sulphuric acid attacks zinc on heating, and sulphur dioxide is liberated:

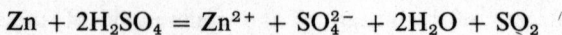

$$Zn + 2H_2SO_4 = Zn^{2+} + SO_4^{2-} + 2H_2O + SO_2$$

(d) *Halogens:* Direct reaction on heating:

$$Zn + Cl_2 = ZnCl_2$$

128

A similar reaction takes place with sulphur, but no reaction occurs with hydrogen or nitrogen.

(e) *Alkalies:* A solution of caustic alkali dissolves zinc on heating:

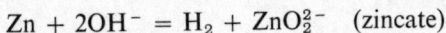

$$Zn + 2OH^- = H_2 + ZnO_2^{2-} \quad \text{(zincate)}$$

Zinc oxide White when cold, and yellow when hot. Used as a pigment in paint (does not blacken in sulphide atmospheres) and in medicaments for skin diseases.

Zinc oxide is amphoteric:

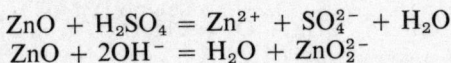

$$ZnO + H_2SO_4 = Zn^{2+} + SO_4^{2-} + H_2O$$
$$ZnO + 2OH^- = H_2O + ZnO_2^{2-}$$

Zinc chloride Very deliquescent, and is soluble in organic solvents. Anhydrous zinc chloride is used as a catalyst in some organic reactions, and is prepared by evaporating a solution of the salt in a stream of hydrogen chloride gas.

Zinc sulphide Precipitated from alkaline solutions of zinc salts by hydrogen sulphide:

$$Zn^{2+} + H_2S = 2H^+ + ZnS$$

Zinc sulphide is a white substance, and is used as a pigment, and in some fluorescent powders.

Zinc complexes Addition of ammonia to zinc solutions produces a solution of the complex tetra-ammine zinc ion:

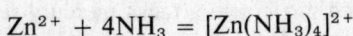

$$Zn^{2+} + 4NH_3 = [Zn(NH_3)_4]^{2+}$$

Outline of the Reactions of Mercury

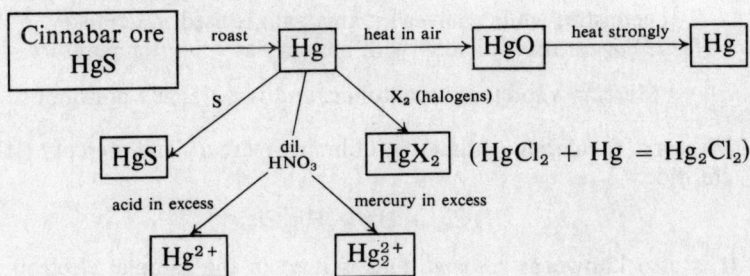

129

Brief Chemistry of Mercury

Ore: mercury (II) sulphide, as cinnabar, HgS.

$E^{\ominus}$ **value:** $+0.85$ V.

Reaction of the metal with:

(a) *Air:* At room temperature, the metallic impurities oxidize gradually and cause the mercury to 'tail'. On heating, mercury forms mercury (II) oxide:

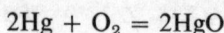

$$2Hg + O_2 = 2HgO$$

On stronger heating, the reaction reverses, and oxygen is liberated.

(b) *Water:* No action.

(c) *Acids:* As would be expected from its position in the electrochemical series, mercury does not liberate hydrogen from acids. Hence it is unaffected by dilute hydrochloric and sulphuric acids. The reaction with hot concentrated sulphuric acid is similar to that of zinc.

With dilute nitric acid, the products depend on the relative quantities of metal and acid present. With excess acid:

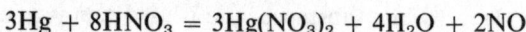

$$3Hg + 8HNO_3 = 3Hg(NO_3)_2 + 4H_2O + 2NO$$

With excess metal:

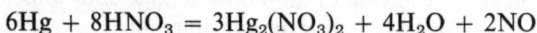

$$6Hg + 8HNO_3 = 3Hg_2(NO_3)_2 + 4H_2O + 2NO$$

Concentrated nitric acid attacks mercury readily, the reaction being similar to that for copper.

(d) *Alkalies:* No action.

(e) *Other metals* often dissolve in mercury to form an amalgam. Sodium amalgam is used as a reducing agent in organic chemistry, while a silver/tin amalgam is used in dentistry.

(f) *Halogens* react directly with mercury at room temperature.

Mercury vapour is monatomic, and is extremely poisonous.

Mercury (I) chloride   Made by subliming mercury and mercury (II) chloride:

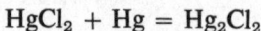

$$HgCl_2 + Hg = Hg_2Cl_2$$

It is also known as *calomel*, and is used in the calomel electrode

130

which is widely employed in connection with electrochemical measurements.

With ammonia, mercury (I) chloride produces mercury and the black ammono-chloride:

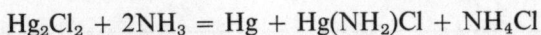

$$Hg_2Cl_2 + 2NH_3 = Hg + Hg(NH_2)Cl + NH_4Cl$$

This reaction is used as a test in the detection of mercury (I) compounds.

Mercury (II) chloride Readily formed by the action of chlorine on mercury:

$$Hg + Cl_2 = HgCl_2$$

It is soluble in water, and is reduced in a two-stage process by tin (II) ions:

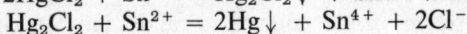

$$2HgCl_2 + Sn^{2+} = Hg_2Cl_2\downarrow + Sn^{4+} + 2Cl^-$$
$$Hg_2Cl_2 + Sn^{2+} = 2Hg\downarrow + Sn^{4+} + 2Cl^-$$

Mercury (II) iodide Precipitated when iodide ion is added to a solution of mercury (II) ion,

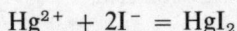

$$Hg^{2+} + 2I^- = HgI_2$$

or by direct combination. It is an orange-red solid, insoluble in water, but soluble in potassium iodide solution:

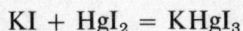

$$KI + HgI_2 = KHgI_3$$

A solution of this complex in the presence of alkali is called *Nessler's solution*, and this solution produces a brown precipitate with traces of ammonia.

10 Aluminium

General Properties

1 Aluminium is a reactive metal forming a tripositive, colourless ion.
2 Aluminium oxide is amphoteric.
3 The small ionic size and the large charge carried by the aluminium ion results in the tendency to form covalent compounds.
4 The charge-to-volume ratio of the aluminium ion accounts for the non-existence of aluminium carbonate, and for the fact that it is not possible to precipitate aluminium sulphide from solution.
5 Aluminium compounds are frequently highly hydrated; for example the sulphates and alums.
6 Aluminium forms some complexes, and many organo-metallic compounds.

Ore: mixed oxide (bauxite), $Al_2O_3.2H_2O$ (together with some iron (III) oxide and silicate material)

$E^\ominus$ **value:** -1.66 V.

Reaction of the metal with:

(a) *Air* (on strong heating),

$$4Al + 3O_2 = 2Al_2O_3 \text{ (and a little AlN)}$$

(b) *Water:* not attacked because of formation of a stable oxide film.
(c) *Acid:* cold dilute hydrochloric,

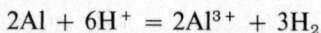

$$2Al + 6H^+ = 2Al^{3+} + 3H_2$$

Dilute sulphuric acid is unable to dissolve the oxide film and, therefore, does not attack aluminium.

Aluminium dissolves in both concentrated hydrochloric and sulphuric acids. Both dilute and concentrated nitric acid render aluminium passive.

132

(d) *Alkali:* warm, dilute alkali solution or hot concentrated alkali metal carbonate solutions dissolve aluminium, to form aluminates:

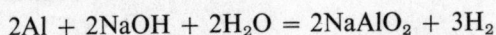

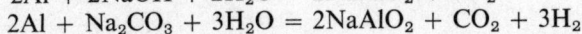

$$2Al + 2NaOH + 2H_2O = 2NaAlO_2 + 3H_2$$
$$2Al + Na_2CO_3 + 3H_2O = 2NaAlO_2 + CO_2 + 3H_2$$

(e) *Metal oxides:* a mixture of aluminium and iron, chromium or manganese oxides reacts when ignited by a burning magnesium wire:

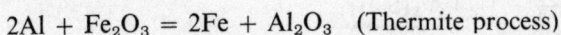

$$2Al + Fe_2O_3 = 2Fe + Al_2O_3 \quad \text{(Thermite process)}$$

(f) *Chlorine:* on heating aluminium in a stream of anhydrous chlorine, anhydrous aluminium chloride is formed:

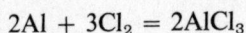

$$2Al + 3Cl_2 = 2AlCl_3$$

This experiment is carried out on a laboratory scale using the apparatus shown in Figure 61. The anhydrous bromide and

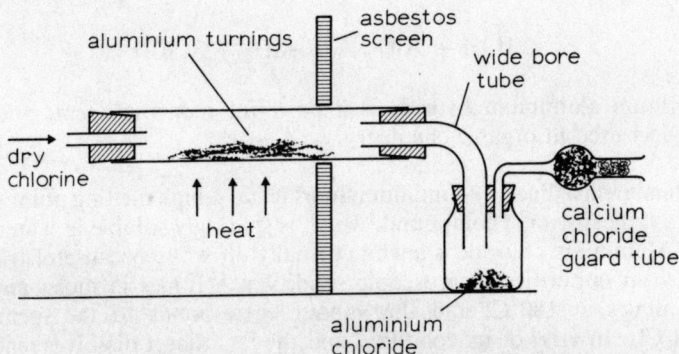

Fig. 61 Preparation of anhydrous aluminium chloride

iodide may be made in a similar manner. Dry hydrogen chloride or hydrogen fluoride may also be used, in which case the product is anhydrous aluminium chloride and fluoride respectively.

133

Outline of the Reactions of Aluminium

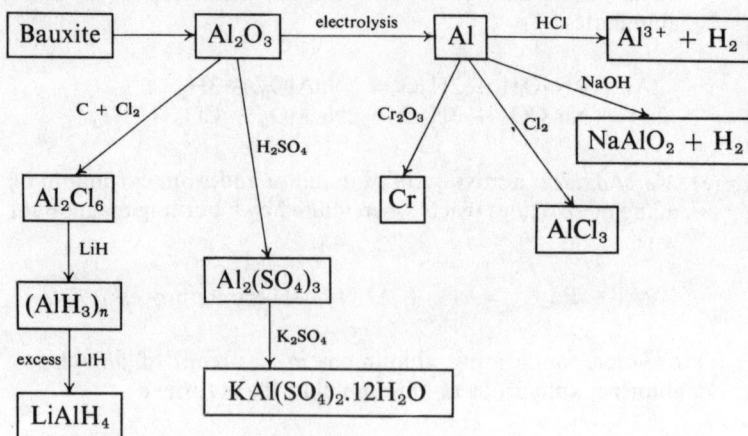

$$\boxed{\text{Bauxite}} \longrightarrow \boxed{Al_2O_3} \xrightarrow{\text{electrolysis}} \boxed{Al} \xrightarrow{\text{HCl}} \boxed{Al^{3+} + H_2}$$

Al → (NaOH) → $\boxed{NaAlO_2 + H_2}$

Al_2O_3 → (C + Cl$_2$) → $\boxed{Al_2Cl_6}$

Al_2O_3 → (H$_2$SO$_4$) → $\boxed{Al_2(SO_4)_3}$

Al → (Cr$_2$O$_3$) → $\boxed{Cr}$

Al → (Cl$_2$) → $\boxed{AlCl_3}$

$\boxed{Al_2Cl_6}$ → (LiH) → $\boxed{(AlH_3)_n}$ → (excess LiH) → $\boxed{LiAlH_4}$

$\boxed{Al_2(SO_4)_3}$ → (K$_2$SO$_4$) → $\boxed{KAl(SO_4)_2.12H_2O}$

Aluminium hydrides When lithium hydride is added to a solution of aluminium chloride in ethoxy-ethane (ether), a white solid of the composition $(AlH_3)_n$ is formed. In the presence of excess lithium hydride, lithium aluminium hydride (which is soluble in ethoxy-ethane) is formed:

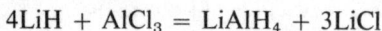

$$4LiH + AlCl_3 = LiAlH_4 + 3LiCl$$

Lithium aluminium hydride is a powerful reducing agent, and is widely used in organic chemistry.

Aluminium halides Aluminium fluoride has a high melting point and is a typically ionic compound, which is sparingly soluble in water.

Aluminium chloride is a white (usually off-white because of traces of iron impurities) hygroscopic solid, which fumes in moist air. It sublimes at 180°C, and the vapour corresponds to the formula Al_2Cl_6. In view of its volatility, and the fact that it dissolves readily in many organic solvents, aluminium chloride is thought to be intermediate between an ionic and a covalent compound. The Al_2Cl_6 molecule has chlorine atoms surrounding each aluminium atom tetrahedrally, with the 'bridging' chlorine atoms each donating a lone pair of electrons to form a coordinate link with each aluminium atom, as indicated in Figure 62.

Anhydrous aluminium chloride is extensively used as a catalyst in organic reactions involving halides; for example, the Friedel-Crafts

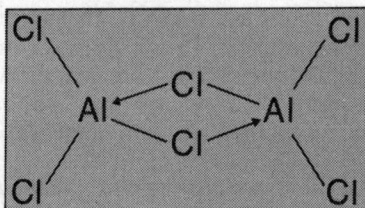

Fig. 62 Structure of aluminium chloride

reaction. When aluminium dissolves in hydrochloric acid, the hydrate $AlCl_3.6H_2O$ is formed. On heating, this salt undergoes hydrolysis, eventually producing aluminium hydroxide or oxide.

Aluminium oxide This is used as a catalyst (as activated alumina), as an abrasive, in the production of synthetic gems (e.g., sapphires), in certain cements and as a furnace lining.

Aluminium oxide is soluble in both acids and alkalies, showing that it is an amphoteric oxide.

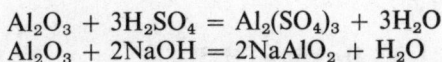

$$Al_2O_3 + 3H_2SO_4 = Al_2(SO_4)_3 + 3H_2O$$
$$Al_2O_3 + 2NaOH = 2NaAlO_2 + H_2O$$

Sodium aluminate solution is used in water-softening, and potassium aluminate can be obtained in a crystalline form.

Aluminium sulphate and the alums Alumina dissolves in hot, concentrated sulphuric acid, forming crystals of the sulphate on evaporation. These crystals are highly hydrated $[Al_2(SO_4)_3.18H_2O]$, but may be rendered anhydrous by careful heating.

Aluminium sulphate is used as a mordant in dyeing. Cloth soaked in aluminium sulphate is steamed, or placed in a dilute alkaline solution, when aluminium hydroxide is formed among the fibres. The aluminium hydroxide readily adsorbs dye during later processes, so that the dye is firmly bound to the cloth.

The addition of aluminium sulphate to a colloidal suspension produces a coagulation and precipitation of the colloidal particles. This is the basis of a method of sewage purification.

The term *alum* is applied to double salts of the type

$$M^IM^{III}(SO_4)_2.12H_2O$$

where M^I is a metal in oxidation state $+1$ (for example sodium, potassium, ammonium) and M^{III} is a metal in oxidation state $+3$ (for example aluminium, chromium, iron (III)).

135

Alums are usually produced by crystallizing equimolar proportions of the two metal sulphates. In some cases, a series of reactions may be chosen which automatically produce the correct proportion of each sulphate in the solution which is prepared for crystallization. For instance, to prepare potassium aluminium alum:

(a) Aluminium turnings are dissolved in potassium hydroxide:

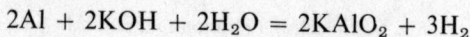

$$2Al + 2KOH + 2H_2O = 2KAlO_2 + 3H_2$$

(b) The solution is then just neutralized with sulphuric acid:

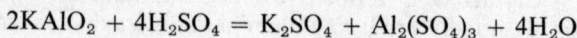

$$2KAlO_2 + 4H_2SO_4 = K_2SO_4 + Al_2(SO_4)_3 + 4H_2O$$

which, on evaporating and crystallizing, forms crystals of potassium aluminium alum, $KAl(SO_4)_2.12H_2O$.

Potassium chromium alum may be produced from the solution resulting from the reaction between acidified potassium chromate (VI) and alcohol:

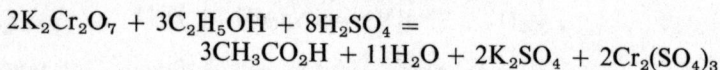

$$2K_2Cr_2O_7 + 3C_2H_5OH + 8H_2SO_4 =$$
$$3CH_3CO_2H + 11H_2O + 2K_2SO_4 + 2Cr_2(SO_4)_3$$

Ammonia alum loses ammonia and sulphuric acid on strong heating, to form a residue of pure alumina.

11 Section 2: The p-Block Elements

THE NON-METALS

General Properties

1 Non-metals are those elements lying to the right of, and above, the diagonal line running from boron to astatine, as shown in the Periodic Table (Table 1, p. 7).

2 Non-metals exhibit such typical properties as:

(a) variable in appearance,
(b) not malleable or ductile,
(c) do not conduct heat or electricity,
(d) form acidic oxides,
(e) form anions by attracting electrons—hence they are the electronegative elements.

3 The electronegativity value of an element can be taken as an indication of the electron-attracting power of an element when in a combined state. The electronegativities of the elements range from 0·7 (caesium) to 4 (fluorine). The values for the commoner non-metals are shown diagrammatically in Figure 63.

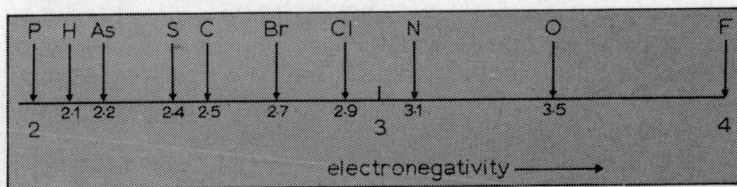

Fig. 63 Electronegativities

(a) Bonds between elements of widely different electronegativities are polar (ionic in the extreme cases).
(b) Bonds between elements of similar electronegativities are covalent, and the polarity is small.

137

(c) Most of the common non-metals have electronegativities between 2 and 3: only the halogens and oxygen participate in anion formation to any significant extent.

(d) Electronegativity values decrease with increasing atomic number in a group. Hence, 'non-metallic' character is most clearly marked in the early members of a group.

(e) Electronegativity values for elements in Group VII are greater than those for the corresponding elements in Group VI, which in turn are greater than the corresponding values in Group V, etc. Thus, non-metallic character decreases down a group in the Periodic Table, but increases from left to right across a period. This makes fluorine the most electronegative element, and also continues the diagonal relationships noted already in connection with the metallic elements (p. 74).

4 The majority of compounds formed between non-metals are covalent, and the chemistry of the non-metals may be discussed in terms of the nature of the covalent bond.

Covalent Bonding

In this type of bonding, a link is formed between two atoms, A and B, when one electron from atom A is shared with one electron from atom B (or when two electrons from either A or B are shared between the two atoms—in this case the link is described as a coordinate covalent bond).

On page 39, this process was described by the merging of two singly filled atomic orbitals inhabited by electrons of opposite spin. As a result of the merging of the atomic orbitals, a molecular orbital is formed, so that one may visualize the atoms as linked in a fixed direction in space. For example, the four hydrogen atoms bonded to carbon in methane are arranged in the form of a regular tetrahedron. An important result of covalent bonding is that the molecules take on a definite shape.

A second fact is that, on the formation of a bond, energy is released. This energy, known as *bond energy*, may be defined as the energy required to split one mole of molecules AB into the component atoms A and B. It is important to realize that, apart from the noble gases, none of the elements exist as single atoms, but are linked to form structural units, ranging from simple diatomic molecules (e.g., Cl_2) to more complex units (e.g., an S_8 ring), and to giant molecules (e.g., diamond). (Examples of these structures are given on pp. 48–50.)

138

The Shapes of Covalent Molecules

The electron pair repulsion theory (Sidgwick–Powell theory) This theory allows the shapes of molecules to be deduced, using the following rules, which simplify handling of the concepts of promotion and hybridization of electrons introduced on page 42.

(a) All the valence electrons in the central atom are involved in determining the shape of the molecule.
(b) To the number of valence electrons present is added the number of electrons contributed for sharing by the various atoms bonded to the central atom.
(c) The total number of electrons so found is divided into pairs.
(d) Each electron pair takes up a position so that the forces of repulsion between the pairs are at a minimum.
(e) A pair of electrons linking an atom to the central atom is known as a *bond pair*.
(f) A pair of electrons residing on the central atom, and not used in bonding, is called a *lone pair*.

Some examples will illustrate the application of these rules.

Methane (CH_4) Carbon has the configuration $1s^2\ 2s^2\ 2p^2$. Thus, total number of valence electrons

(= number in quantum group 2) =	4
Four hydrogen atoms contribute	4
Total	8 = 4 pairs

The pairs of electrons repel each other so that they are as far apart as possible; that is they form a regular tetrahedron. Each pair of electrons is used in linking a hydrogen atom to the central carbon atom. All the electron pairs are bond pairs.

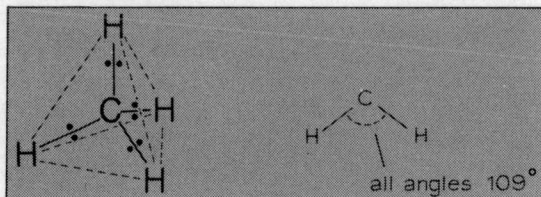

Fig. 64 Methane molecule

139

Ammonia (NH_3) Nitrogen, configuration $1s^2\,2s^2\,2p^3$, contributes five valence electrons; three hydrogen atoms contribute three. This gives a total of eight, that is four pairs.

Of these four pairs, three are bond pairs, and there is one lone pair. Because of mutual repulsion, a tetrahedral arrangement is produced. However, the lone pair of electrons is held by the nitrogen atom only and is located close to this atom, while the bonding pairs

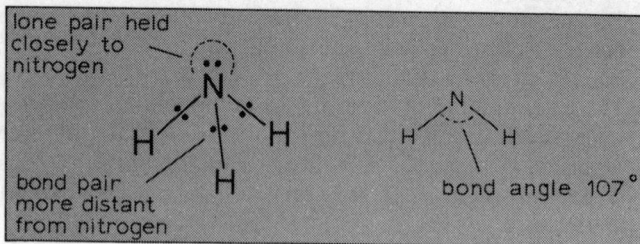

Fig. 65 Ammonia molecule

of electrons lie further away between the nitrogen and hydrogen atoms. This results in a displacement of the bonding pairs towards each other, and the interbond angle falls from 109° to 107°.

NF_3 and NCl_3 have similar structures.

Water (H_2O) The number of valence electrons on the central atom (oxygen) is six, and the two hydrogen atoms contribute two electrons. Thus the total is eight, that is four pairs. Two of these pairs are bonding pairs, and two are lone pairs.

Again, the arrangement is basically a tetrahedral displacement of electron pairs, but the concentration of the two lone pairs close to the oxygen atom displaces the two bonding pairs from a regular arrangement, so that the inter-bond angle falls further to 104°

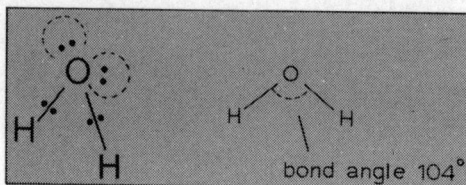

Fig. 66 Water molecule

140

Sulphur hexafluoride (SF_6)   The number of valence electrons on the central (sulphur) atom is six, and the six fluorine atoms contribute another six electrons. Thus the total is twelve, that is six pairs.

The six bonding pairs take up a regular octahedral arrangement, and the interbond angle is 90°

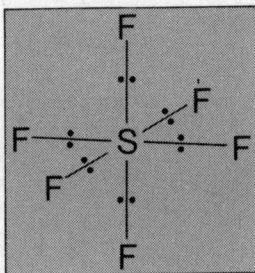

Fig. 67 Sulphur hexafluoride

Phosphorus pentachloride (PCl_5)   Phosphorus has a total of five valence electrons. Five chlorine atoms add five extra electrons, giving a total of ten, that is five pairs.

A perfectly regular arrangement resulting from the equal repulsion of five pairs of electrons cannot be established. The structure which approximates most closely to a regular arrangement is a *trigonal bipyramid*. The angle between the bond pair above (or below) the plane and the bond pairs in the plane is 90°. The angle between the bond pairs in the plane is 120°. Consequently, the mutual repulsion of the bond pairs within the plane is less than the mutual repulsion of these bond pairs and those above and below the plane. This leads to a slight difference in the P–Cl distances, as shown in Figure 68.

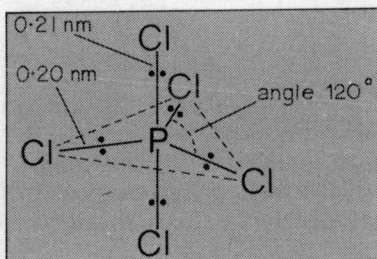

Fig. 68 Phosphorus pentachloride

141

Chlorine trifluoride (ClF_3) The chlorine atom, with structure $1s^2 \, 2s^2 \, 2p^6 \, 3s^2 \, 3p^5$, contributes seven valence electrons. Three fluorine atoms contribute three electrons, giving a total of ten. Of these five pairs of electrons, three are bond pairs, and two are lone pairs. The basic arrangement is again a trigonal bipyramid, with two lone pairs in the equatorial plane.

The resulting molecule is 'T' shaped, with an inter-bond angle of $87\frac{1}{2}°$, as shown in Figure 69.

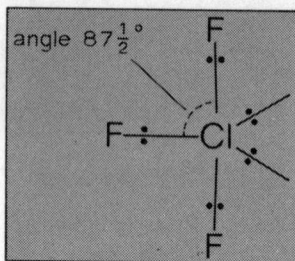

Fig. 69 Chlorine trifluoride

The structures of these and other covalent molecules which can be deduced using the above theory are summarized in Table 14, page 143.

Covalent Bond Energies

A reaction between covalent molecules involves:

(a) breaking the bonds uniting the atoms of the reactant molecules,
(b) remaking bonds to produce molecules of products.

For example:

435 kJ of energy is needed to dissociate 1 mole of hydrogen molecules to form 2 moles of hydrogen atoms. Similarly, 242 kJ of energy is required to produce 2 moles of chlorine atoms from 1 mole of chlorine gas. Thus, the total energy required to form 1 mole of hydrogen and 1 mole of chlorine atoms from the respective molecules is

$$217·5 + 121 = 338·5 \text{ kJ}$$

142

The bond energy of hydrogen chloride is 430 kJ. Hence, when the reaction

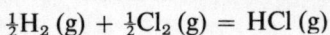

$$\tfrac{1}{2}H_2\,(g) + \tfrac{1}{2}Cl_2\,(g) = HCl\,(g)$$

takes place, $338\cdot5 - 430 = -91\cdot5$ kJ of energy is transferred. (The negative sign means that energy is released.)

TABLE 14 *The shapes of some covalent molecules*

Total number of electron pairs	Basic arrangement	Example	Number of bond pairs	Number of lone pairs	Actual shape of molecule
2	linear	$BeCl_2$	2	0	Cl—Be—Cl linear
3	plane triangular	BCl_3	3	0	Cl | B ∕ \\ Cl Cl equilateral triangle
		$SnCl_2$	2	1	Sn ∕ \\ Cl Cl 'V' shaped
4	tetrahedral	CH_4	4	0	H | C ∕ | \\ H H H tetrahedral
		NH_3	3	1	N ∕ | \\ H H H pyramidal
		H_2O	2	2	O ∕ \\ H H 'V' shaped

143

TABLE 14 (continued)

Total number of electron pairs	Basic arrangement	Example	Number of bond pairs	Number of lone pairs	Actual shape of molecule
5	trigonal bipyramid	PCl_5	5	0	trigonal bipyramid
		SF_4	4	1	distorted tetrahedron
		ClF_3	3	2	'T' shaped
6	octahedral	SF_6	6	0	octahedral
		IF_5	5	1	square pyramid
		XeF_4	4	2	square planar

This quantity of energy is the *heat of formation* of hydrogen chloride, and it represents the *balance* between the large amount of energy required to break the bonds linking the reactant atoms, and that released when the product molecules are formed. In consequence, the heat of formation does not give a direct indication of the strength of the bonds involved.

The hydrogen–chlorine bond is a strong link, needing 430 kJ per mole to break it. This explains why hydrogen chloride appears as the product in many reactions in which these two elements are present, and also why hydrogen chloride is seldom decomposed during a reaction.

Fluorine is a particularly reactive element. This is due largely to two factors:

(a) The bond strength of the fluorine molecule is small.
(b) The strength of a bond linking fluorine to most other elements is high.

Data showing the energies required to break 1 mole of single bonds linking two atoms are shown for a number of elements in Table 15.

TABLE 15 *Some single bond energies (kJ/mol)*

	H	C	N	S	I	Br	Cl	F
H	435	415	389	339	297	364	430	564
C	415	343	293	259	238	272	326	439
Cl	430	326	200	255	213	217	242	255
F	564	439	272	297		255	255	158

Some typical conclusions that may be drawn from these energy values are:

(a) The low bond energies linking fluorine or chlorine atoms in their respective molecules agrees with the high reactivity of these elements.
(b) The high bond strength of the carbon–hydrogen bond, and the much lower bond strengths of the C-Cl, C-Br and C-I links, agree with the unreactivity of the hydrocarbons and the greater reactivity of the halogeno–alkanes.

145

(c) The very high bond strength of the carbon–fluorine link explains the inertness of the fluorinated hydrocarbons.

(d) The low bond strength of the N–Cl bond agrees with the instability of nitrogen trichloride.

The Manner of Rupture of Covalent Bonds

A covalent bond is identified with a shared pair of electrons. On breakage, the bond may rupture in one of two ways:

(a) Each atom retains one electron:

$$A:B \rightarrow A \cdot + \cdot B$$

This is called *homolytic* fission of a bond, and the products are atoms (or free radicals if either A or B is a group rather than an atom).

This type of rupture is usually initiated by the absorption of energy, such as heat or light energy, while a chain reaction may result. For example:

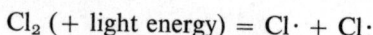

$$Cl_2 \ (+ \ \text{light energy}) = Cl \cdot + Cl \cdot$$

Then

$$Cl \cdot + H_2 = HCl + H \cdot$$
$$H \cdot + Cl_2 = HCl + Cl \cdot \quad \text{etc.}$$

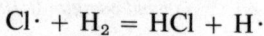

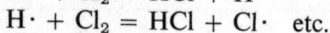

As light energy is continually fed into the system, the formation of atoms increases rapidly, and the rate of reaction increases to give an explosion.

(b) One atom takes both electrons from the bond:

$$A:B = A^- + B^+$$

This is termed *heterolytic* fission of a bond, and the products are ions. This type of bond rupture is influenced by polar solvents which can supply the energy needed to break further bonds by solvation of the resulting ions.

The ionization of hydrogen chloride in the presence of water is an example of this type of bond fission:

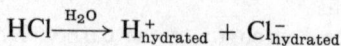

$$HCl \xrightarrow{H_2O} H^+_{\text{hydrated}} + Cl^-_{\text{hydrated}}$$

Covalent molecules most likely to undergo this type of fission are those in which the bond is already fairly polar.

146

The Noble Gases

Discovery   Argon was discovered in 1894 by Rayleigh and Ramsey, who were investigating the discrepancy between the density of nitrogen prepared by the normal laboratory method, and that of atmospheric nitrogen (produced by sparking out all the 'reactive' constituents of air). Helium was discovered in connection with the spectroscopic examination of the chromosphere of the sun during a total eclipse. Following this, Ramsay suggested that these newly discovered elements were two members of a new group (Group 0) in the Periodic Table, and that three elements still remained undiscovered. In 1898, krypton and xenon were found in the residues resulting from the evaporation of liquid air, while neon was discovered in a sample of liquid argon. Radon is radioactive, with a half-life of 4 days (for the longest-lived isotope).

Uses   Argon is used mainly for a protective inert atmosphere in the welding of reactive metals, and in other cases where an inert atmosphere is essential. All the gases are used in discharge tubes and in illuminated signs.

Isolation   Helium is present in the natural gas from American oil wells, and it is isolated from that source. Argon is the most abundant noble gas, and is present in the atmosphere to the extent of 0·9 per cent. It is produced by the fractionation of liquid air, and is commercially available compressed in cylinders. Neon is also obtained from the fractionation of liquid air, but it is far less abundant than argon.

Properties

The noble gases are monatomic (i.e., they exist as single atoms) and have, therefore, low boiling and melting points:

	m.p.	b.p.
Helium	$-272°C$	$-269°C$
Neon	-249	-246
Argon	-189	-186
Krypton	-157	-153
Xenon	-112	-107

They are colourless and odourless and were initially called *rare* gases on account of the difficulty of isolating them. When argon and neon were prepared in commercial quantities they were renamed the *inert* gases, since they appeared not to undergo chemical combination.

However, Bartlett, in 1962, treated platinum hexafluoride with excess xenon at room temperature, and found that a reaction took place, in which a yellow solid, $Xe^+(PtF_6)^-$, was produced. Since this discovery, many compounds of xenon, and a few of krypton, have been made, so that the reluctance of these gases to react, rather than their complete inertness, is reflected in their present collective name of *noble* gases.

Compounds of Xenon

Heating fluorine in a metal vessel with an excess of xenon produces the tetrafluoride:

$$Xe + 2F_2 = XeF_4$$

By heating xenon and excess fluorine under pressure, the hexafluoride is formed:

$$Xe + 3F_2 = XeF_6$$

The tetrafluoride is a colourless, crystalline solid, which sublimes easily. The hexafluoride is similar, but is much more volatile. Both compounds react with hydrogen:

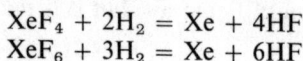

$$XeF_4 + 2H_2 = Xe + 4HF$$
$$XeF_6 + 3H_2 = Xe + 6HF$$

The quantity of hydrogen fluoride produced in these reactions is used to indicate the formula of the compound, or, alternatively, for the quantitative estimation of the xenon fluoride present.

Both fluorides can be hydrolysed to produce XeO_3, while xenon hexafluoride attacks silica:

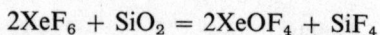

$$2XeF_6 + SiO_2 = 2XeOF_4 + SiF_4$$

Xenon difluoride, krypton difluoride and krypton tetrafluoride have also been prepared.

12 The Chemistry of the Non-Metals

HYDROGEN

The special properties displayed by hydrogen makes the assignment of the element to a particular group in the Periodic Table impossible. Hydrogen stands apart from the rest of the elements in that:

1 It can form a hydride ion H^- (oxidation state -1) when in combination with a highly electropositive metal.
2 It can show an oxidation state of $+1$ (as H^+, or rather H_3O^+) in aqueous solution.
3 It has an electronegativity value of $2 \cdot 1$ (which is close to that of carbon), and can take part in covalent bonding. Since the electronic configuration is $1s^1$, a single bond produced by electron sharing is formed.
4 It is able to form a second weaker bond (i.e., a hydrogen bond) in conjunction with a highly electronegative atom, for example, oxygen, nitrogen or fluorine.

Hydrogen, in fact, appears in more compounds than any other element. Water is the most abundant compound of hydrogen. Natural gas, crude oil and coal all contain combined hydrogen, mostly in the form of hydrocarbons.

Small-Scale Preparation

1 Action of a metal standing above hydrogen in the electrochemical series, on dilute acid or water:

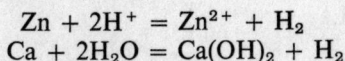

$$Zn + 2H^+ = Zn^{2+} + H_2$$
$$Ca + 2H_2O = Ca(OH)_2 + H_2$$

2 Reaction of metal hydrides with water:

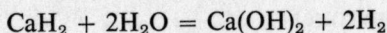

$$CaH_2 + 2H_2O = Ca(OH)_2 + 2H_2$$

149

3 Electrolysis of dilute acids or alkalies. With acids, hydrogen ions are discharged at the cathode:

$$2H^+ + 2e^- = H_2$$

With alkalies, for example, sodium hydroxide, both hydrogen and sodium ions are present at the cathode. In general, the ion lowest in the electrochemical series accepts electrons (and is, therefore, discharged) most readily. Hence, for example:

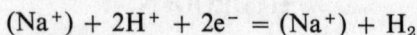

$$(Na^+) + 2H^+ + 2e^- = (Na^+) + H_2$$

Manufacture   Hydrogen is produced industrially as follows:

1 *The water gas reaction.* Steam is passed through very hot coke; the gaseous products are known as *water gas:*

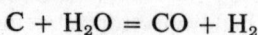

$$C + H_2O = CO + H_2$$

Water gas will reduce a further quantity of steam when passed over an iron (II) oxide catalyst at 500°C:

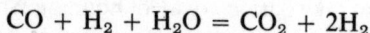

$$CO + H_2 + H_2O = CO_2 + 2H_2$$

The carbon dioxide is removed by washing with hot potassium carbonate solution under pressure:

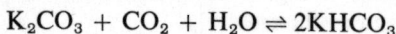

$$K_2CO_3 + CO_2 + H_2O \rightleftharpoons 2KHCO_3$$

2 Recently, the production of hydrogen from hydrocarbons (natural gas and by-products from petroleum refining) has become an important process, and this method is displacing the water gas reaction.

Steam is injected into the hydrocarbon mixture heated to 700°C. In the presence of a nickel catalyst, the product is a mixture of hydrogen, methane, carbon monoxide and carbon dioxide. A further reaction in the presence of iron (II) (and/or other metal oxides) converts the methane to hydrogen and carbon dioxide. The carbon dioxide is removed by potassium carbonate solution, as in method 1.

Uses (a) Synthesis of chemicals such as hydrogen chloride, methanol and ammonia.
  (b) Hydrogenation of edible oils to form solid fats (margarine manufacture).

(c) Reduction of certain metal oxides.

(d) As a rocket fuel.

Ortho and Para Hydrogen

Hydrogen is a diatomic molecule, and the two protons may spin in parallel directions (both clockwise, for example) or in opposite directions. The former is known as *ortho* hydrogen, the latter as *para*. These two forms are in equilibrium, and at low temperatures, the *para* form predominates. At room temperature, hydrogen is a mixture containing 75 per cent *ortho* (the maximum possible concentration), and sudden changes in temperature do not bring about a

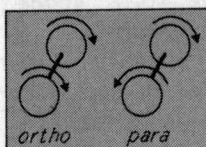

Fig. 70 *Ortho* and *para* hydrogen

rapid change in the relative concentrations unless a catalyst is present. When *ortho* changes to the *para* form, heat is evolved. Hence during the production of liquid hydrogen, the liquid is passed over a catalyst to speed up the decrease in the *ortho-para* ratio; if this were not done, much liquid hydrogen would be boiled away by the heat produced in the conversion from *ortho* to *para* (no matter how efficient the thermal insulation of the vessel) during subsequent storage.

The two forms can be detected as they show slightly different physical properties such as freezing point, boiling point and thermal conductivity.

Deuterium and Tritium

Deuterium (symbol D) and tritium (symbol T) are isotopes of hydrogen, of relative atomic masses 2 and 3 respectively. Thus, the nucleus of deuterium contains one proton and one neutron, and that of tritium contains one proton and two neutrons. Deuterium is found in ordinary hydrogen to the extent of 0·015 per cent. Tritium is radioactive (half-life 12·5 years) and it is produced by a nuclear reaction.

Deuterium and hydrogen show different physical properties, while corresponding differences are found between water and deuterium oxide (heavy water), as shown in Table 16.

151

TABLE 16

	Hydrogen	Deuterium	Water	Deuterium oxide
Density at 4°C			1·000	1·107 g/cm³
Boiling point	−252·6°	−249·5°	100°	101·4°C
Freezing point	−259·1°	−254·3°	0°	3·8°C
Specific heat			1·000	1·018
Latent heat of vaporization	0·92	1·27	4·06	4·17 kJ/mol

Deuterium is produced by the repeated electrolysis of sodium hydroxide dissolved in heavy water. Initially, little deuterium is discharged; later, the gas evolved at the cathode contains a proportion of deuterium, and is burnt to form water, which is recycled. Successive electrolyses eventually yield nearly pure deuterium.

Water may be enriched in deuterium oxide by the *exchange process*. This is based on the fact that the hydrogen in some compounds (e.g., H_2S) can exchange with the deuterium in water:

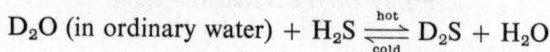

$$D_2O \text{ (in ordinary water)} + H_2S \underset{\text{cold}}{\overset{\text{hot}}{\rightleftharpoons}} D_2S + H_2O$$

The process is shown in Figure 71. Hydrogen sulphide is passed in counter-current to a stream of ordinary hot water. The D_2S formed by exchange is retrieved by the water in the cold static water vessel, and the hydrogen sulphide is recycled. In this way, the concentration of deuterium oxide in the static water builds up to 2 per cent. Fractional distillation increases the concentration to over 90 per cent.

Heavy water is used as a moderator in nuclear reactors, and as a non-radioactive isotope in tracer studies.

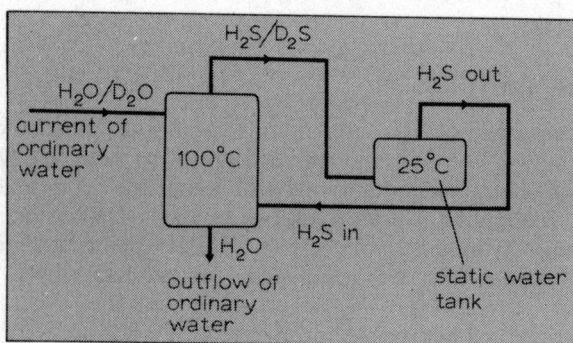

Fig. 71 Concentration of heavy water by the exchange process

Outline of the Reactions of Hydrogen

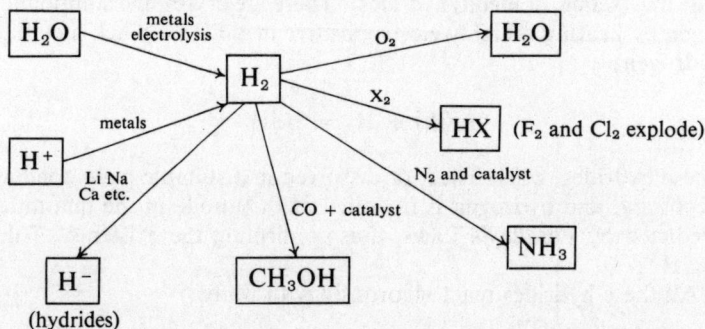

Reaction of hydrogen with:

(a) *Oxygen:* Burns with a pale blue flame:

$$2H_2 + O_2 = 2H_2O$$

Ignition of a hydrogen–oxygen mixture usually produces an explosion.

(b) *Halogens:* Fluorine explodes spontaneously when mixed with hydrogen even in the dark:

$$F_2 + H_2 = 2HF$$

Chlorine combines with hydrogen, often explosively on exposure to heat or light:

$$H_2 + Cl_2 = 2HCl$$

although hydrogen may be burnt in chlorine, or vice versa. Bromine and iodine combine only on heating, and the reaction with iodine is reversible:

$$H_2 + I_2 \rightleftharpoons 2HI$$

(c) *Transition metals:* Palladium and platinum adsorb or *occlude* large quantities of hydrogen gas. On heating, the occluded hydrogen is released, and it is probable that the adsorbed hydrogen is in some form of loose combination with the metal.

153

Types of Hydride

Salt-like (saline, or metal) hydrides   These are crystalline compounds made by heating a highly electropositive metal (an s-block metal) in hydrogen;

$$2Li + H_2 = 2LiH$$

These hydrides, when fused, or dissolved in a suitable melt, conduct electricity, and hydrogen is liberated at the anode in the quantities predicted by Faraday's Laws, thus confirming the existence of the ion H^-.

All these hydrides react vigorously with water:

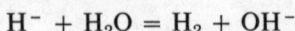

$$H^- + H_2O = H_2 + OH^-$$

Covalent hydrides   These are the hydrides of boron, and the non-metallic elements. The hydrogen atom is joined to the central non-metal atom in these compounds by single covalent bonds, to produce discrete molecules of low relative molecular mass; thus these hydrides are alternatively classed as gaseous, molecular hydrides. Many of these compounds will be met with in later sections concerned with the individual non-metals. Methods of producing these hydrides vary according to the non-metal present, but the following route is generally applicable to the preparation of covalent hydrides.

A compound produced by the combination of a non-metal and a highly electropositive metal yields the corresponding non-metal hydride when reacted with hydrogen in the $+1$ oxidation state (i.e., water or acid). For example:

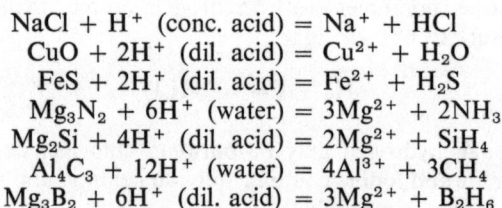

$$NaCl + H^+ \text{ (conc. acid)} = Na^+ + HCl$$
$$CuO + 2H^+ \text{ (dil. acid)} = Cu^{2+} + H_2O$$
$$FeS + 2H^+ \text{ (dil. acid)} = Fe^{2+} + H_2S$$
$$Mg_3N_2 + 6H^+ \text{ (water)} = 3Mg^{2+} + 2NH_3$$
$$Mg_2Si + 4H^+ \text{ (dil. acid)} = 2Mg^{2+} + SiH_4$$
$$Al_4C_3 + 12H^+ \text{ (water)} = 4Al^{3+} + 3CH_4$$
$$Mg_3B_2 + 6H^+ \text{ (dil. acid)} = 3Mg^{2+} + B_2H_6$$

The hydrides of boron form a special class of covalent hydrides. Diborane is produced by the last reaction given above, but a better method is to reduce boron trifluoride with lithium hydride in dry ethoxy-ethane (ether):

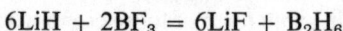

$$6LiH + 2BF_3 = 6LiF + B_2H_6$$

BH_3 is unknown. Diborane is a reactive gas, which inflames in moist air, and is readily hydrolysed to boric acid and hydrogen. Until recently, the electronic structure of diborane and other related hydrides has been obscure. The molecule is now known to have the structure in which the hydrogen atoms linking the two boron atoms

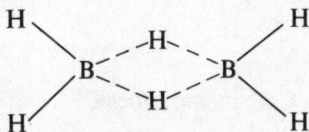

form a unique bond known as a *bridging bond*. Altogether, the four bonds between the two bridging hydrogen atoms and the two boron atoms contain a total of four electrons. These bonds are thus distinct from the normal electron pair covalent bond, and the structure is said to be *electron deficient*.

Complex hydrides Two important complex hydrides, lithium aluminium hydride and sodium borohydride, are widely used as reducing agents in organic chemistry. They are made as follows:

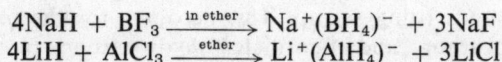

$$4NaH + BF_3 \xrightarrow{\text{in ether}} Na^+(BH_4)^- + 3NaF$$
$$4LiH + AlCl_3 \xrightarrow{\text{ether}} Li^+(AlH_4)^- + 3LiCl$$

Although these compounds do not contain free hydride ions, they behave in a similar way to the salt-like hydrides, except that the reactions take place more smoothly.

Other hydrides Many metals, especially transition metals, react with hydrogen to form hydrides in which the ratio of metal atoms to hydrogen atoms is not one of simple whole numbers. Examples are $ThH_{3.1}$ and $CeH_{2.7}$. Such compounds are said to be *non-stoichiometric*. These hydrides resemble metals rather than salts, as many of the properties of the metal appear unchanged. However, heat is liberated during their formation, so that some degree of combination is indicated, although its exact nature is not clear.

GROUP IVB: CARBON, SILICON, TIN AND LEAD

General Properties

1 The elements in Group IVB show a transition from non-metallic to metallic as the atomic number increases. There is a decrease in electronegativity from 2·5 for carbon to 1·4 for lead.

2 All the elements can show a covalency of four.
3 As the atomic number increases, there is a tendency to show an ionic valency of $+2$. This corresponds to a reluctance of two of the four valence electrons to enter into combination.
4 All the elements can form volatile hydrides, and volatile alkyl derivatives.

Properties Specific to Carbon and Silicon

1 Carbon can form strong links with itself (C–C bond strength 343 kJ per mole) whereas the silicon–silicon bond is weaker (200 kJ per mole).
2 Only carbon–hydrogen and carbon–fluorine bonds are stronger than the carbon–carbon bond. Silicon forms bonds with many elements which are stronger than the silicon–silicon bond. Thus, a chain of carbon atoms is stable, but a chain of silicon atoms is less stable. Also, a reaction leading to the formation of a bond between carbon and another element by breaking carbon–carbon bonds leads (except in the case of hydrogen and fluorine) to weaker bonds being set up. This accounts for the inert nature of hydrocarbon and fluorinated hydrocarbon molecules. A corresponding reaction with silicon leads, in almost every case, to the formation of stronger bonds, as follows:

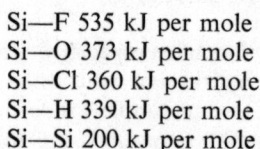

$$
\begin{aligned}
&\text{Si—F } 535 \text{ kJ per mole} \\
&\text{Si—O } 373 \text{ kJ per mole} \\
&\text{Si—Cl } 360 \text{ kJ per mole} \\
&\text{Si—H } 339 \text{ kJ per mole} \\
&\text{Si—Si } 200 \text{ kJ per mole}
\end{aligned}
$$

Thus, in many reactions of silicon compounds, products containing silicon–halogen or silicon–oxygen bonds are formed.
3 Carbon has the electronic structure $1s^2\,2s^2\,2p^2$, and after forming four covalent bonds by electron sharing, the element cannot take part in any further bonding.

Silicon is $1s^2\,2s^2\,2p^6\,3s^2\,3p^2\,(3d^0)$. After forming four covalencies, silicon can use the empty 3d-orbitals (which are valence level orbitals) to accept further pairs of electrons. In this way, it is possible to increase the extent of bonding to a maximum of six;

$$SiF_4 + 2HF = H_2[SiF_6]$$

The ion has the structure

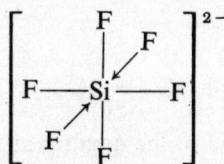

$$\left[\begin{array}{c} F \\ | \quad F \\ F-Si-F \\ F \quad | \\ F \end{array} \right]^{2-}$$

4 Carbon has a much smaller atomic radius than silicon and can take part in multiple bond formation, which it exhibits in bonds between itself, oxygen and nitrogen. Silicon is a larger atom, and does not show this property.

(*Note:* Hydrogen is rather similar to carbon in that it has a corresponding 'half filled' valence level, and a similar electronegativity.)

CARBON

The two allotropes of carbon are diamond and graphite. The structures and properties of these two allotropes were outlined on pages 49–50.

Carbon Monoxide

The gas is prepared by dehydrating methanoic or ethanedioc acid (or their salts) with hot, concentrated sulphuric acid:

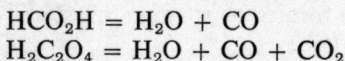

$$HCO_2H = H_2O + CO$$
$$H_2C_2O_4 = H_2O + CO + CO_2$$

The carbon dioxide is removed by bubbling the gas stream through sodium hydroxide solution.

Carbon monoxide is a colourless, odourless and poisonous gas. It is stable to heat, and sparingly soluble in water. It is regarded as a neutral oxide, although it will combine with hot concentrated alkali to produce a methanoate:

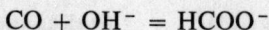

$$CO + OH^- = HCOO^-$$

The gas burns in air with a pale blue flame:

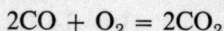

$$2CO + O_2 = 2CO_2$$

157

It dissolves in a solution of copper (I) chloride in hydrochloric acid.

Carbon monoxide is a strong reducing agent, and reduces many oxides:

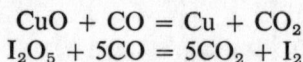

$$CuO + CO = Cu + CO_2$$
$$I_2O_5 + 5CO = 5CO_2 + I_2$$

This latter reaction is used in the quantitative estimation of carbon monoxide.

In sunlight, or in the presence of a charcoal catalyst, carbon monoxide combines with chlorine to give another poisonous gas, carbon dichloride oxide, or phosgene:

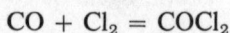

$$CO + Cl_2 = COCl_2$$

A special type of reaction is the formation of metal carbonyls by the reaction of carbon monoxide with a transition metal. The easiest to prepare is nickel carbonyl:

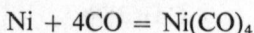

$$Ni + 4CO = Ni(CO)_4$$

This is made by passing carbon monoxide over finely divided nickel at 30°C. Nickel carbonyl decomposes on further heating, and this method is used for the production of pure nickel.

Carbonyls are also used as starting points for the preparation of many organo-metallic complexes of the transition elements.

Carbon Dioxide

The gas occurs in the atmosphere, and is produced on a commercial scale from the gases formed on roasting limestone, or from fermentation processes.

In the laboratory, it is made by the action of dilute acid on a carbonate:

$$2H^+ + CO_3^{2-} = H_2O + CO_2$$

or by heating sodium hydrogen carbonate:

$$2NaHCO_3 = Na_2CO_3 + H_2O + CO_2$$

Properties Carbon dioxide is a colourless, odourless gas, heavier than air, and having a slight acid reaction to moist litmus. It dissolves in water to produce a solution of the weak carbonic acid:

$$CO_2 + H_2O = H_2CO_3$$

158

Carbon dioxide does not burn, nor does it support combustion.

When the gas is bubbled through calcium hydroxide solution, a precipitate of calcium carbonate forms:

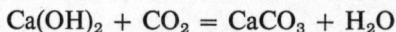

$$Ca(OH)_2 + CO_2 = CaCO_3 + H_2O$$

On prolonged passage of the gas, the more soluble calcium hydrogen carbonate is produced, and the initial cloudiness of the suspension almost disappears:

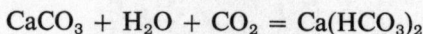

$$CaCO_3 + H_2O + CO_2 = Ca(HCO_3)_2$$

This reaction explains the presence of temporary hardness in water. On boiling, carbon dioxide is expelled and a precipitate forms:

$$Ca(HCO_3)_2 = CaCO_3 + CO_2 + H_2O$$

Carbonates

1 All carbonates, except those of the alkali metals (lithium is an exception) and ammonium, are insoluble.

2 Addition of a soluble carbonate to a solution of a soluble metal salt usually gives a precipitate of the basic carbonate, for example, $CuCO_3 . Cu(OH)_2$. The reason for this is that the carbonate ion undergoes hydrolysis:

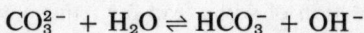

$$CO_3^{2-} + H_2O \rightleftharpoons HCO_3^- + OH^-$$

and the hydroxide ions also form a precipitate. To overcome this action, precipitation is usually effected by adding sodium or potassium hydrogen carbonate to the metal salt solution, as the presence of the HCO_3^- ions greatly reduces the hydroxide ion concentration.

3 All carbonates are decomposed by dilute acids, liberating carbon dioxide:

$$CO_3^{2-} + 2H^+ = CO_2 + H_2O$$

4 All the insoluble carbonates decompose on heating:

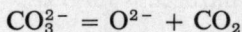

$$CO_3^{2-} = O^{2-} + CO_2$$

The carbonate ion is a large planar ion, and the electron screen surrounding this ion is easily distorted by the close presence of small positive ions which carry a high charge. Such cations produce a strong polarizing effect, and the carbonate ion decomposes on heating. An extreme example of this effect accounts for the non-existence of aluminium carbonate.

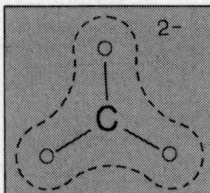

Fig. 72 Carbonate ion

The Manufacture and Use of some Carbon Compounds

Carbon dioxide is manufactured by roasting limestone in air, and from alcoholic fermentation processes. Another source of the gas is the by-product from the water gas reaction.

Carbon dioxide is used:

(a) In the manufacture of ammonium sulphate from powdered anhydrite:

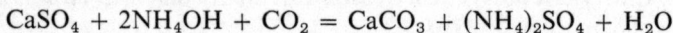

$$CaSO_4 + 2NH_4OH + CO_2 = CaCO_3 + (NH_4)_2SO_4 + H_2O$$

(b) In the production of white lead for paint pigments.

(c) For the production of urea, by heating with ammonia under pressure at 200°C:

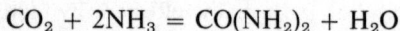

$$CO_2 + 2NH_3 = CO(NH_2)_2 + H_2O$$

Urea is used as a fertilizer and in the manufacture of plastics.

Other uses of carbon dioxide include the manufacture of organic chemicals, as a coolant in nuclear reactors, for the aeration of soft drinks, for storing fruit; while blocks of solid carbon dioxide are used as a refrigerant.

Carbon monoxide is manufactured in the form of *producer gas*. Air is passed through a bed of white-hot coke, and a mixture of carbon monoxide and nitrogen is formed:

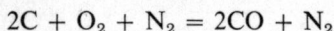

$$2C + O_2 + N_2 = 2CO + N_2$$

Some carbon dioxide is also present, and the gas is used as a fuel.

160

Carbon disulphide is made by the reaction between methane and sulphur vapour at a high temperature and in the presence of a catalyst:

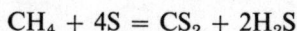

$$CH_4 + 4S = CS_2 + 2H_2S$$

Carbon disulphide is widely used in the manufacture of viscose rayon and cellophane.

Tetrachloromethane is manufactured by the reaction of chlorine and carbon disulphide:

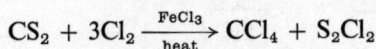

$$CS_2 + 3Cl_2 \xrightarrow[\text{heat}]{\text{FeCl}_3} CCl_4 + S_2Cl_2$$

Also,

$$CS_2 + 2S_2Cl_2 \xrightarrow[\text{heat}]{\text{FeCl}_3} CCl_4 + 6S$$

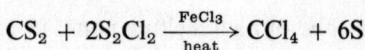

Tetrachloromethane is used as a dry-cleaning fluid, and in fire extinguishers. However, on a hot metal surface, in air, tetrachloromethane can produce carbon dichloride oxide, and for this reason the use of tetrachloromethane for the above purposes is diminishing. It is used as a starting material for the manufacture of modern refrigerant gases such as 'freon' and 'arcton'.

SILICON

Silicon occurs naturally as silica, SiO_2, and it is prepared by high temperature reduction with magnesium or carbon:

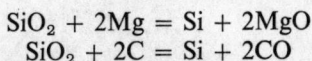

$$SiO_2 + 2Mg = Si + 2MgO$$
$$SiO_2 + 2C = Si + 2CO$$

Very pure silicon is produced by the reduction of silicon tetrachloride or a chlorosilane with zinc at 1000°C:

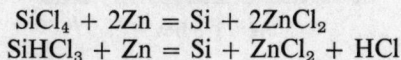

$$SiCl_4 + 2Zn = Si + 2ZnCl_2$$
$$SiHCl_3 + Zn = Si + ZnCl_2 + HCl$$

The silicon is further purified by zone refining, and then forms the starting material for the manufacture of semi-conductor devices.

Outline of the Reactions of Silicon

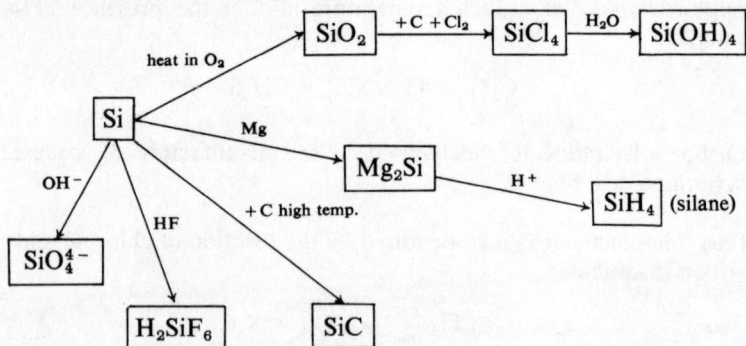

Silica

Silicon dioxide is a solid of high melting point, in contrast to carbon dioxide which is a gas at s.t.p. The difference in physical properties is due to the structural differences between the compounds. Carbon can form multiple bonds with oxygen, and the molecule of carbon dioxide is linear, $O{=}C{=}O$.

Silicon forms four bonds, displaced tetrahedrally about the silicon atom, and each of these is linked to an oxygen atom, producing a giant molecule. The structure of this is shown (distorted into a two-dimensional representation) in Figure 73. In this diagram, the repeating unit is the (encircled) SiO_2 unit.

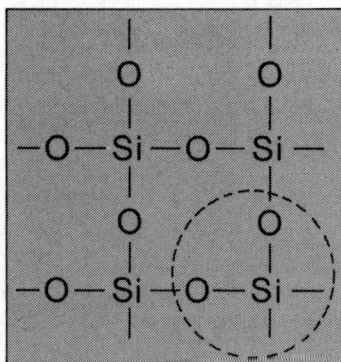

Fig. 73 Silica

Silica in the form of quartz has a very low coefficient of expansion, and it can withstand high temperatures. Hence, it is used in the manufacture of special apparatus, for example, the envelopes for water-cooled high-current rectifying valves. Sand (impure silica) is used for making mortar, concrete and in glass manufacture.

Silicates

When silicon tetrachloride reacts with water, orthosilicic acid is produced:

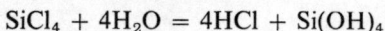

$$SiCl_4 + 4H_2O = 4HCl + Si(OH)_4$$

This reaction can continue by way of the successive elimination of the elements of water from orthosilicic acid, finally producing silica:

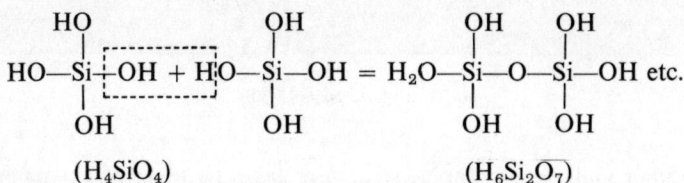

HO   OH   OH   OH
 | | | |
HO—Si┼OH + H┊O—Si—OH = H₂O—Si—O—Si—OH etc.
 | | | |
OH   OH   OH   OH

(H_4SiO_4)   ($H_6Si_2O_7$)

The successive acids are known as *condensed* acids, the term implying that the product results from the elimination of the elements of water between two reactant molecules. (See Part 2.)

The *orthosilicate* ion, SiO_4^{4-}, may be regarded as the fundamental unit from which the various silicate structures are built up. It is a tetrahedral ion, but it can be represented by the triangle shown in Figure 74(b), which is the view obtained by looking down on to the tetrahedron directly from above. In orthosilicates, the anions are independent, and the cations (usually bipositive) are present with these ions in a regular lattice. An example is Mg_2SiO_4 (olivine).

By linking two orthosilicate tetrahedra, more complex structures result. Figure 75 shows the unit formed when the units are linked

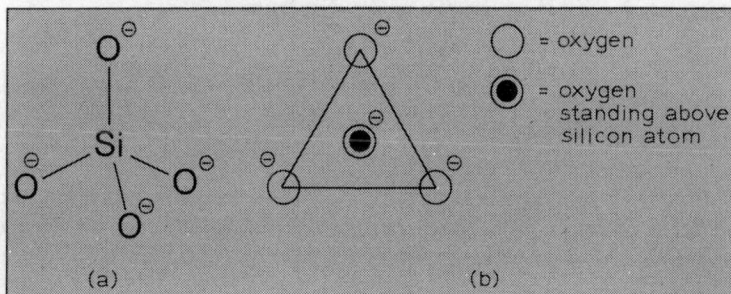

Fig. 74 Orthosilicate ion, SiO_4^{4-}

163

by way of a common oxygen atom. Note that the linking oxygen does not now carry a charge. The condensed ion is $Si_2O_7^{6-}$, and an example is $Sc_2(Si_2O_7)$ (thorvetite).

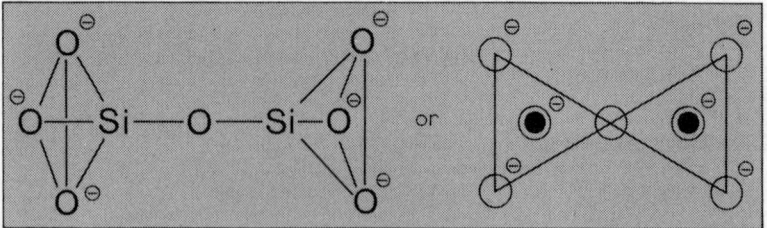

Fig. 75 $Si_2O_7^{6-}$ ion

Other and more complex structures take the form of chains and sheets. Figure 76 shows the single chain or *pyroxene* structure. The formula of the repeating unit is $(SiO_3)^{2-}$, often called the *metasilicate* ion. An example is $CaMg(SiO_3)_2$ (diopside).

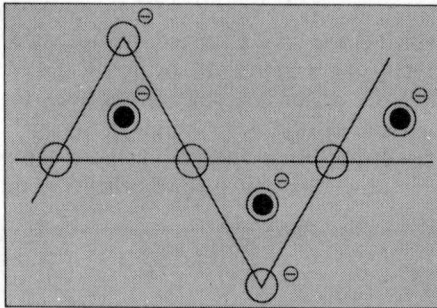

Fig. 76 Pyroxene

Figure 77 shows a sheet or *amphibole* structure, which is a linking to form double chains. Both pyroxenes and amphiboles are easily split into strands. Further linkages are possible, and these lead to the formation of a large two-dimensional network, or sheet-like structures (e.g., mica), and eventually the three-dimensional framework of silica is reached.

164

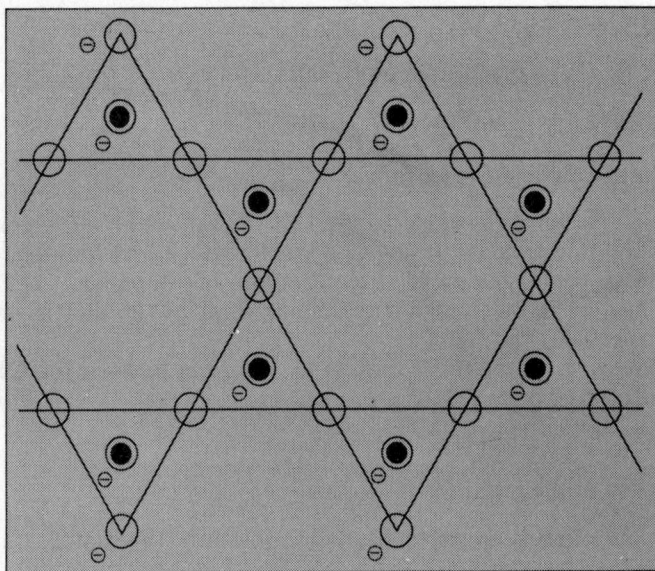

Fig. 77 Amphibole

TIN AND LEAD

Properties Specific to Tin and Lead

1 Group IV as a whole shows a complete change from typically non-metallic (carbon) to metallic properties (with tin and lead), although the metals are not highly electropositive.
2 Weak metallic bonding is evident from the malleability, ductility and low melting points and boiling points of these metals.
3 Both metals can show both a $+2$ and a $+4$ oxidation state. With tin, the latter state is preferred; with lead, the former.
4 Tin and lead dioxides are amphoteric, although lead dioxide is less acidic, and is an oxidizing agent.
5 Both metals form organo-metallic compounds.

Outline of the Reactions of Tin

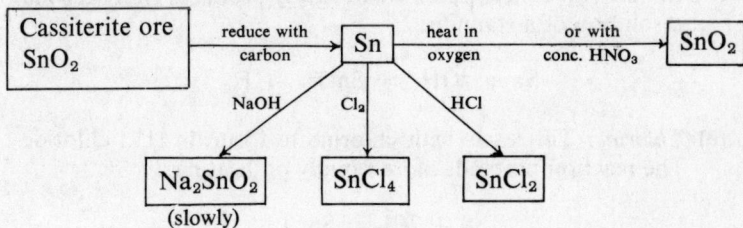

Brief Chemistry of Tin

Ore: tin (IV) oxide as cassiterite, SnO_2

$E^\ominus$ **value:** -0.14 V (Sn^{2+}/Sn)

Tin has three *allotropic forms:*

1 *Grey tin* which is stable below 13°C. When white tin is cooled below this temperature, it slowly falls into a grey powder, the change being most rapid at temperatures of -40°C to -50°C.
2 *White tin* is the common crystalline form of tin, and is stable between 13° and 162°C.
3 *Rhombic tin* is the stable form at temperatures between 162°C and the melting point (232°C).

Reaction of the metal with:

(a) *Air:* Reacts on strong heating to form tin (IV) oxide:

$$Sn + O_2 = SnO_2$$

(b) *Water:* Tin decomposes steam when very strongly heated:

$$Sn + 2H_2O = SnO_2 + 2H_2$$

(c) *Acids:* Very slow reaction with dilute hydrochloric and nitric acid. With concentrated hydrochloric acid:

$$Sn + 2HCl = SnCl_2 + H_2$$

Hot concentrated sulphuric acid gives a reaction similar to zinc. Concentrated nitric acid in the presence of air produces a white powder which is mainly hydrated tin (IV) oxide (meta-stannic acid).

(d) *Alkali:* Hot concentrated alkali slowly produces hydrogen and a solution of a stannite:

$$Sn + 2OH^- = SnO_2^{2-} + H_2$$

(e) *Chlorine:* Tin reacts with chlorine to form tin (IV) chloride. The reaction proceeds more rapidly on heating.

$$Sn + 2Cl_2 = SnCl_4$$

Outline of the Reactions of Lead

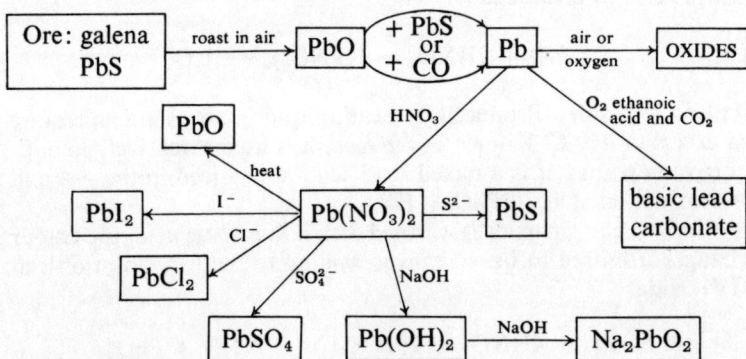

Brief Chemistry of Lead

Ore: lead sulphide, as galena, PbS

$E^{\ominus}$ **value:** -0.13 V

Reaction of the metal with:

(a) *Air:* Produces yellow lead (II) oxide on heating:

$$2Pb + O_2 = 2PbO$$

On heating in air above 450°C, trilead tetroxide is formed:

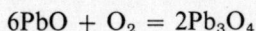

$$6PbO + O_2 = 2Pb_3O_4$$

(b) *Water:* Soft water dissolves lead gradually as $Pb(OH)_2$.
(c) *Acids:* Hydrochloric acid has little effect on lead. Concentrated sulphuric acid dissolves lead on heating, while nitric acid attacks lead when both dilute and concentrated. The reactions are similar to those given for copper.
(d) *Alkalies:* They have no effect on lead.

The Oxides of Lead

Lead (II) oxide Produced either by heating lead in air, or by heating lead nitrate:

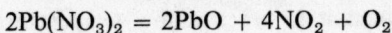

$$2Pb(NO_3)_2 = 2PbO + 4NO_2 + O_2$$

167

Lead (II) oxide is a basic oxide which reacts with dilute acid (e.g., nitric acid) to form lead (II) salts:

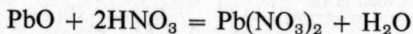

$$PbO + 2HNO_3 = Pb(NO_3)_2 + H_2O$$

Trilead tetroxide   Produced by heating lead in air at a temperature in excess of 450°C. Known as *red lead*, it is used extensively in anti-corrosion paints. It is a mixed oxide containing lead in the $+4$ and $+2$ oxidation states $(2PbO + PbO_2)$.

When trilead tetroxide is warmed with dilute nitric acid, the colour changes from red to brown corresponding to the formation of lead (IV) oxide:

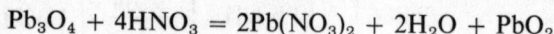

$$Pb_3O_4 + 4HNO_3 = 2Pb(NO_3)_2 + 2H_2O + PbO_2$$

Lead (IV) oxide   A higher oxide and a strong oxidizing agent. With concentrated hydrochloric acid:

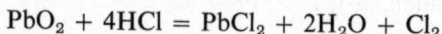

$$PbO_2 + 4HCl = PbCl_2 + 2H_2O + Cl_2$$

Tetra-ethyl lead   Used as an anti-knock agent in petrol engines. It is manufactured by the reaction of a sodium lead alloy with chloroethane:

$$4NaPb + 4C_2H_5Cl = 3Pb + 4NaCl + (C_2H_5)_4Pb$$

Lead tetra-ethanoate   Made by dissolving trilead tetroxide in glacial ethanoic acid. It is hydrolysed by water, and is a powerful oxidizing agent. It is used for this purpose in organic chemistry.

GROUP VB: NITROGEN, PHOSPHORUS, ARSENIC, ANTIMONY, AND BISMUTH

General Properties

1 The non-metallic nature of these elements decreases with increasing atomic number. Nitrogen can form multiple bonds, and hence exists as a diatomic molecule, $N \equiv N$. Phosphorus and arsenic form tetrahedral structural units, P_4 and As_4, while antimony and bismuth form metallic linkages.
2 Phosphorus, arsenic and antimony exhibit allotropy, each having a non-metallic and a more metallic form.
3 Nitrogen is reluctant to react at low temperatures, but it does

168

react with metals such as lithium or magnesium on heating. Phosphorus is more reactive than nitrogen at lower temperatures, combining readily with oxygen, chlorine and a number of metals. For the remaining elements in the group, the reactivity shows a steady decline.

4 Anions bearing a triple negative charge are unstable in solution. The main chemistry of these elements is the chemistry of covalent compounds. All the elements can show a covalency of three; for example, they all form a hydride, EH_3 (but BiH_3 has only been made in trace quantities).

5 Triply covalent nitrogen has a lone pair of electrons which can be used in the formation of a coordinate link (e.g., NH_4^+). The maximum covalency of nitrogen is thus four. The remaining elements can use all their valency electrons in bonding, so giving a maximum covalency of five.

6 The decrease in non-metallic character with increase in atomic number, allows bismuth to exhibit metallic properties in an oxidation state $+3$.

7 The oxides of the elements are usually acidic. Dinitrogen oxide (N_2O) is usually regarded as a neutral oxide, while bismuth (III) oxide (Bi_2O_3) is basic.

NITROGEN

The gas is obtained from the atmosphere by liquefaction and fractionation. It is used (in liquid form) as a refrigerant (it is safer than liquid air) and as an inert atmosphere. Ammonia is manufactured from atmospheric nitrogen.

In the laboratory, nitrogen is produced by heating a solution containing ammonium and nitrite ions:

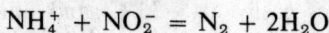

$$NH_4^+ + NO_2^- = N_2 + 2H_2O$$

Nitrogen is reluctant to react with air, water, acids, alkalies and most metals and non-metals at low or moderate temperatures. Lithium and magnesium however, form nitrides on heating:

$$6Li + N_2 = 2Li_3N$$

Ammonia

On the laboratory scale, ammonia is displaced from ammonium salts on heating with an alkali:

$$NH_4^+ + OH^- = NH_3 + H_2O$$

169

The gas is collected by upward delivery, and may be dried over calcium oxide.

Ammonia is manufactured by the Haber process, in which nitrogen (obtained from the atmosphere) is mixed with hydrogen (obtained from the petroleum industry) and reacted under pressure in the presence of a finely divided iron catalyst:

$$N_2 + 3H_2 \rightleftharpoons 2NH_3$$

The pure hydrogen and nitrogen mixture is compressed to some 300 atmospheres, and about 15 per cent of the mixture is converted to ammonia. The ammonia is removed by condensing out, while the unchanged gases are continually recycled through the catalyst chamber.

Uses (a) A great deal is converted into fertilizer, for example, ammonium sulphate.

      (b) Nitric acid is manufactured by the catalytic oxidation of ammonia

      (c) As a refrigerant.

      (d) In the production of polyamide plastics, and many organic chemicals.

Properties and Reactions

Ammonia is a colourless gas, with a pungent smell. It is lighter than air, toxic, and very soluble in water, producing an alkaline solution:

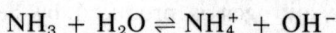

$$NH_3 + H_2O \rightleftharpoons NH_4^+ + OH^-$$

1 Ammonia liquefies easily at $-33°C$, and liquid ammonia is a good solvent. Alkali and alkaline earth metals dissolve in liquid ammonia to give blue solutions. No hydrogen is evolved, and these solutions are thought to contain solvated electrons:

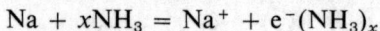

$$Na + xNH_3 = Na^+ + e^-(NH_3)_x$$

These solutions are powerful reducing agents, and conduct electricity almost as well as metals. On heating, or in the presence of iron ions, (as catalyst), amides are formed. For instance, sodamide is formed thus:

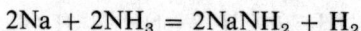

$$2Na + 2NH_3 = 2NaNH_2 + H_2$$

170

These amides are violently attacked by water:

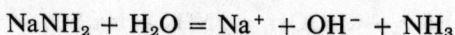

$$NaNH_2 + H_2O = Na^+ + OH^- + NH_3$$

2 Ammonia burns in oxygen when heated:

$$4NH_3 + 3O_2 = 2N_2 + 6H_2O$$

However, in the presence of a platinum catalyst, nitrogen oxide is formed:

$$4NH_3 + 5O_2 = 4NO + 6H_2O$$

3 Ammonia reduces metal oxides:

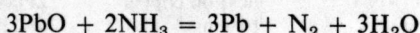

$$3PbO + 2NH_3 = 3Pb + N_2 + 3H_2O$$

4 With chlorine, ammonia being in excess:

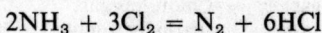

$$2NH_3 + 3Cl_2 = N_2 + 6HCl$$

then,

$$NH_3 + HCl = NH_4Cl$$

With chlorine in excess; nitrogen trichloride, which is a strongly explosive gas, is formed:

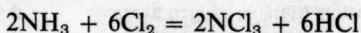

$$2NH_3 + 6Cl_2 = 2NCl_3 + 6HCl$$

5 With solutions of transition metal ions, complexes may be formed:

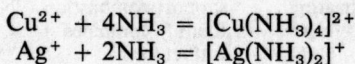

$$Cu^{2+} + 4NH_3 = [Cu(NH_3)_4]^{2+}$$
$$Ag^+ + 2NH_3 = [Ag(NH_3)_2]^+$$

6 A solution of ammonia in water precipitates metal hydroxides:

$$Fe^{3+} + 3NH_4OH = 3NH_4^+ + Fe(OH)_3$$

Ammonium salts The ammonium ion, NH_4^+, can function as a unipositive cation in salts. All ammonium salts are soluble, and dissociate or decompose on heating:

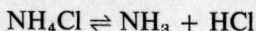

$$NH_4Cl \rightleftharpoons NH_3 + HCl$$

171

Ammonium salts produce ammonia gas on heating with an alkali:

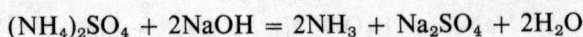

$$(NH_4)_2SO_4 + 2NaOH = 2NH_3 + Na_2SO_4 + 2H_2O$$

Oxides of Nitrogen

The properties of the common oxides of nitrogen are summarized in Table 17.

TABLE 17 *The three common oxides of nitrogen*

	Dinitrogen oxide N_2O	*Nitrogen oxide* NO	*Nitrogen dioxide* $NO_2(N_2O_4)$
Preparation	Heat ammonium nitrate: NH_4NO_3 $= N_2O + 2H_2O$	Action of dil. nitric acid on copper turnings: $3Cu + 8HNO_3$ $= 3Cu(NO_3)_2$ $+ 4H_2O + 2NO$	Heat lead nitrate: $2Pb(NO_3)_2$ $= 2PbO + O_2$ $+ 2N_2O_4$
Physical properties	Colourless gas; faint sweet taste; fairly soluble in water; neutral to litmus	Colourless gas; sparingly soluble in water. Freely soluble in iron (II) sulphate solution, forming $FeSO_4 . NO$, from which pure NO evolved on heating	Brown, poisonous gas, condenses to a pale yellow liquid (when pure) on cooling to $20°$. Choking smell, and soluble in water producing nitrous and nitric acids: $2NO_2 + H_2O$ $= HNO_3 + HNO_2$
Chemical properties	Supports combustion better than air, since it decomposes on heating: $2N_2O = 2N_2 + O_2$ Reacts with sodamide on heating: $NaNH_2 + N_2O$ $= H_2O + NaN_3$ Sodium azide is produced	Combines with oxygen at room temperature: $2NO + O_2 = 2NO_2$ Stable on heating and does not support combustion. Readily combines with chlorine: $2NO + Cl_2$ $= 2NOCl$ nitrogen chloride oxide is formed	Undergoes thermal dissociation: N_2O_4 (colourless solid) $\Updownarrow -10°$ N_2O_4 (colourless liquid) $\Updownarrow 20°$ N_2O_4 (colourless gas) $\Updownarrow 140°$ $2NO_2$ (dark brown gas) $\Updownarrow 620°$ $2NO + O_2$ (colourless mixture) Reduced by carbon monoxide, copper and hydrogen: $CO + NO_2$ $= CO_2 + NO$ $2NO_2 + 7H_2$ $= 2NH_3 + 4H_2O$ $4Cu + 2NO_2$ $= 4CuO + N_2$

172

Nitrous Acid

A dilute solution of the acid is prepared by acidification of a nitrite:

$$KNO_2 + H^+ = K^+ + HNO_2$$

It is unstable in all but cold, dilute aqueous solutions, decomposing according to the equation:

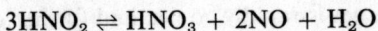

$$3HNO_2 \rightleftharpoons HNO_3 + 2NO + H_2O$$

Nitrous acid and nitrites can function as reducing agents:

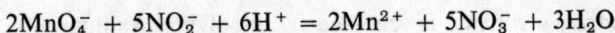

$$2MnO_4^- + 5NO_2^- + 6H^+ = 2Mn^{2+} + 5NO_3^- + 3H_2O$$

and as oxidizing agents:

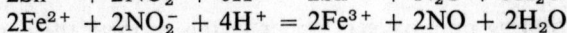

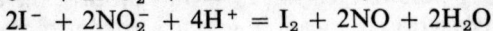

$$2Sn^{2+} + 2NO_2^- + 6H^+ = 2Sn^{4+} + N_2O + 3H_2O$$
$$2Fe^{2+} + 2NO_2^- + 4H^+ = 2Fe^{3+} + 2NO + 2H_2O$$
$$2I^- + 2NO_2^- + 4H^+ = I_2 + 2NO + 2H_2O$$

The liberation of iodine from iodide ion is often used as a test for nitrous acid, although it must be remembered that other oxidizing agents can behave in a similar way.

A special reaction of nitrous acid is the formation of diazonium compounds (see Part 2):

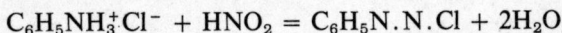

$$C_6H_5NH_3^+Cl^- + HNO_2 = C_6H_5N.N.Cl + 2H_2O$$

Nitric Acid

This acid is prepared in the laboratory by the action of hot, concentrated sulphuric acid on a nitrate:

$$NO_3^- + H_2SO_4 = HSO_4^- + HNO_3$$

The nitric acid is often decomposed by the heat of the reaction, and the nitrogen dioxide formed dissolves in the acid, colouring it yellow:

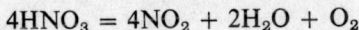

$$4HNO_3 = 4NO_2 + 2H_2O + O_2$$

The acid may be concentrated by distillation with concentrated sulphuric acid. The pure acid, which is difficult to make, is colourless, and is dissociated even at room temperature:

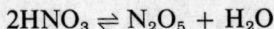

$$2HNO_3 \rightleftharpoons N_2O_5 + H_2O$$

173

Manufacture On a large scale, nitric acid is produced by the catalytic oxidation of ammonia. Dry ammonia gas, mixed with excess air, is heated, under pressure, to 500°C, and passed rapidly over a platinum gauze catalyst. A 90 per cent conversion of the feed is obtained:

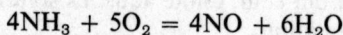

$$4NH_3 + 5O_2 = 4NO + 6H_2O$$

The gaseous product is rapidly cooled and further oxidized:

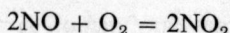

$$2NO + O_2 = 2NO_2$$

The nitrogen dioxide is removed by dissolving in dilute nitric acid solution, the reaction being expressed by the equation:

$$3NO_2 + H_2O = 2HNO_3 + NO$$

By substituting pure oxygen for air in this process, the conversion rate is greatly increased; steam is injected to control the reaction, and this leads directly to the production of nitric acid.

Reactions of Nitric Acid

1 Pure dry nitric acid is a covalent substance, and has no action on metals or carbonates. In aqueous solution, ionization occurs:

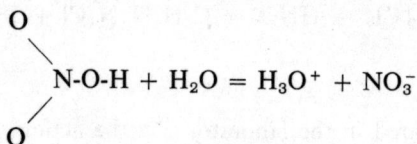

$$\begin{array}{c} O \\ \diagdown \\ N\text{-}O\text{-}H + H_2O = H_3O^+ + NO_3^- \\ \diagup \\ O \end{array}$$

and it behaves as a strong monobasic acid. The salts of nitric acid, the nitrates, are produced by the action of nitric acid on the metal, metal oxide, hydroxide or carbonate; for example,

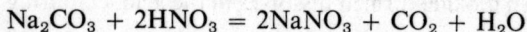

$$Na_2CO_3 + 2HNO_3 = 2NaNO_3 + CO_2 + H_2O$$

2 Hot, concentrated nitric acid is a powerful oxidizing agent, nitrogen oxide usually being produced, although the actual products vary with the conditions. For example, iodine is oxidized to iodic acid:

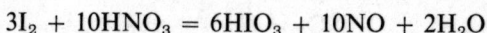

$$3I_2 + 10HNO_3 = 6HIO_3 + 10NO + 2H_2O$$

174

and iron (II) is oxidized to iron (III):

$$3Fe^{2+} + HNO_3 + 3H^+ = 3Fe^{3+} + NO + 2H_2O$$

The nitrogen oxide liberated in this reaction combines with the iron (II) ions to form a complex ion $[Fe(H_2O)_5NO]^{2+}$; this is the species responsible for the colour produced in the brown ring test.

Phosphorus and sulphur are oxidized to phosphoric and sulphuric acids respectively.

3 In the presence of concentrated sulphuric acid, the nitronium ion NO_2^+ is formed from concentrated nitric acid:

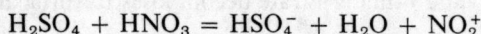

$$H_2SO_4 + HNO_3 = HSO_4^- + H_2O + NO_2^+$$

This ion is involved in the nitration of hydrocarbons:

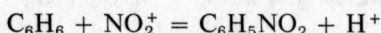

$$C_6H_6 + NO_2^+ = C_6H_5NO_2 + H^+$$

The Action of Nitric Acid on Metals

Nearly all metals react with nitric acid, but aluminium is not attacked to any significant extent, while iron and chromium are made passive by the acid.

The products of the reaction of nitric acid and a metal depend on the conditions under which the reaction takes place:

(a) Very dilute nitric acid does not oxidize the hydrogen liberated.

$$Mg + 2H^+ = Mg^{2+} + H_2$$

This may be regarded as the primary reaction that takes place when a metal is attacked by nitric acid. The way in which the nitric acid oxidizes the hydrogen produced determines the final products of the reaction.

(b) Dilute nitric acid reacts with copper as follows:

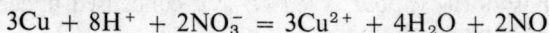

$$3Cu + 8H^+ + 2NO_3^- = 3Cu^{2+} + 4H_2O + 2NO$$

(c) With concentrated nitric acid:

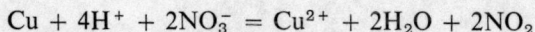

$$Cu + 4H^+ + 2NO_3^- = Cu^{2+} + 2H_2O + 2NO_2$$

Metal Nitrates

All nitrates are soluble. The nitrate ion is a planar ion, and, as in the carbonate ion, the electron screen is easily distorted by the close approach of a small cation carrying a high charge. Consequently, metal nitrates decompose on heating:

$$2M(NO_3)_2 = 2MO + 4NO_2 + O_2$$

Except for lithium, alkali metal nitrates do not decompose in this way (alkali metal ions are large, and carry a single positive charge only, so that they are not strongly polarizing):

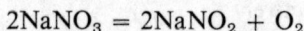

$$2NaNO_3 = 2NaNO_2 + O_2$$

Lithium has a small ion, and is more polarizing than the other alkali metals, and lithium nitrate decomposes to form the oxide on heating:

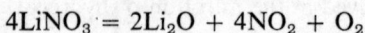

$$4LiNO_3 = 2Li_2O + 4NO_2 + O_2$$

Ammonium nitrate produces dinitrogen oxide on heating:

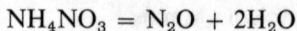

$$NH_4NO_3 = N_2O + 2H_2O$$

PHOSPHORUS

Phosphorus is found in phosphate deposits, for example, apatite $CaF_2.3Ca_3(PO_4)_2$ or calcium phosphate $Ca_3(PO_4)_2$. Most of the phosphate minerals are used for fertilizers.

Phosphorus is extracted from these sources by mixing the ore with silica sand and crushed coke, and heating in an electric furnace. The reactions are usually described by the following equations:

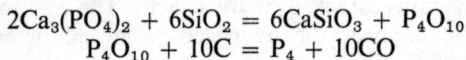

$$2Ca_3(PO_4)_2 + 6SiO_2 = 6CaSiO_3 + P_4O_{10}$$
$$P_4O_{10} + 10C = P_4 + 10CO$$

The calcium silicate slag is tapped from the base of the furnace, and the phosphorus vapour is led off and condensed under water. The product is white phosphorus.

Allotropes

Phosphorus has three main allotropes—white (or yellow) phosphorus, red phosphorus and black phosphorus. Black phosphorus resembles graphite in both appearance and properties. It is made by heating white phosphorus under high pressure. It is the least stable allotrope, and changes to the red form on heating above 550°C.

176

TABLE 18 *Properties of white and red phosphorus*

	White phosphorus	Red phosphorus
Specific gravity	1·83	2·2–2·3
M.P.	44·1°C	592·5°C (under pressure)
Odour	Resembles garlic Poisonous	None Non-poisonous
Solubility	Insoluble in water, soluble in carbon disulphide	Insoluble in both water and carbon disulphide
With alkali	Reacts to give phosphine	No action
In air	Phosphoresces and oxidizes. Catches fire at 50°C	No action. Burns when heated above 240°C
Chlorine	Immediate reaction	Reacts only on warming

The structure of white phosphorus is shown on page 49. Red phosphorus has a polymeric type of structure resulting from the partial breakdown and interlinking of the P_4 tetrahedra.

White phosphorus is metastable under all conditions, and changes slowly into the red form. The change is effected more rapidly by heating in an inert atmosphere, with a small quantity of iodine as a catalyst.

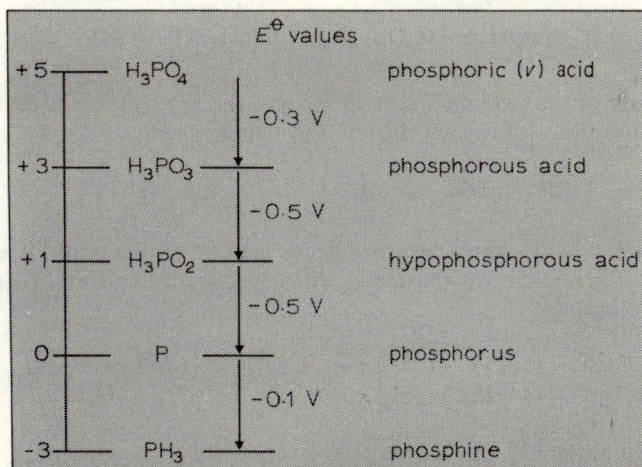

Fig. 78 Oxidation state chart for phosphorus

From this chart, phosphorus in all its oxidation states (except $+5$) is a reducing species. The most stable state is the $+5$ state, and, in most reactions, phosphorus tends to achieve a higher oxidation state. In alkaline solutions, the reducing action is more pronounced.

Outline of the Reactions of Phosphorus

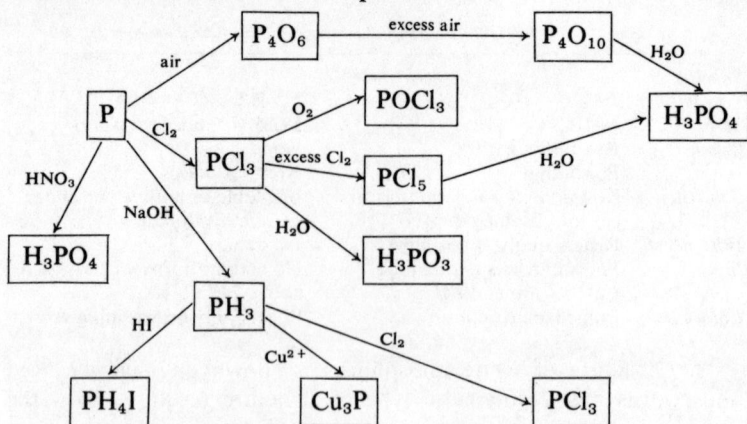

Phosphine, PH_3

Phosphine is a colourless gas, with an unpleasant odour of bad fish. It is inflammable in air, and is prepared by the action of sodium hydroxide solution on white phosphorus.

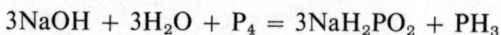

$$3NaOH + 3H_2O + P_4 = 3NaH_2PO_2 + PH_3$$

Because of the inflammability of the gas, the apparatus is filled with an atmosphere of carbon dioxide throughout the experiment.

$$4PH_3 + 8O_2 = P_4O_{10} + 6H_2O \quad \text{(i.e., } 4H_3PO_4)$$

Phosphine is a strong reducing agent, and forms the phosphide, or gives a precipitate of the metal when passed through many metal salt solutions:

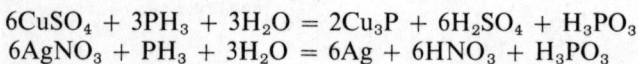

$$6CuSO_4 + 3PH_3 + 3H_2O = 2Cu_3P + 6H_2SO_4 + H_3PO_3$$
$$6AgNO_3 + PH_3 + 3H_2O = 6Ag + 6HNO_3 + H_3PO_3$$

Unlike ammonia, phosphine is not very soluble in water, and it is only feebly basic. The best-known phosphonium compound is phosphonium iodide:

$$PH_3 + HI = PH_4I$$

Again, in contrast to ammonia, phosphine does not behave as a ligand in forming complexes with transition metal ions.

Phosphorus Chlorides, PCl_3 and PCl_5

Phosphorus trichloride is made by passing a stream of dry chlorine over red or white phosphorus.

$$P_4 + 6Cl_2 = 4PCl_3$$

If the phosphorus is kept in excess, the product is mainly the trichloride, and this is distilled (to free it from any pentachloride) from the product mixture.

Phosphorus trichloride is a liquid of boiling point 73°C, and readily hydrolysed by water:

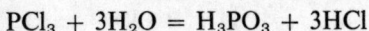

$$PCl_3 + 3H_2O = H_3PO_3 + 3HCl$$

(During hydrolysis, the oxidation state of the phosphorus remains unchanged.)

Phosphorus pentachloride is an off-white solid, made by the direct combination of the elements with the chlorine in excess, or by the chlorination of the trichloride:

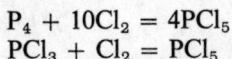

$$P_4 + 10Cl_2 = 4PCl_5$$
$$PCl_3 + Cl_2 = PCl_5$$

On heating to 150°C, dissociation takes place:

$$PCl_5 \rightleftharpoons PCl_3 + Cl_2$$

The structure of the phosphorus pentachloride molecule in the vapour state is a trigonal bipyramid (see p. 141) but in the solid state, it appears to have a rock salt type of lattice containing the ions

$$[PCl_4]^+ \quad [PCl_6]^-$$

Phosphorus pentachloride hydrolyses readily:

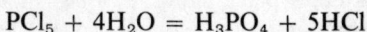

$$PCl_5 + 4H_2O = H_3PO_4 + 5HCl$$

and it reacts with 'OH' groups in both organic and inorganic compounds:

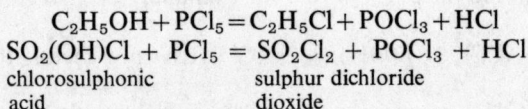

$$C_2H_5OH + PCl_5 = C_2H_5Cl + POCl_3 + HCl$$
$$SO_2(OH)Cl + PCl_5 = SO_2Cl_2 + POCl_3 + HCl$$

chlorosulphonic   sulphur dichloride
acid   dioxide

179

The products of the reaction of phosphorus pentachloride with water may be deduced from a knowledge of electronegativity values. Chlorine is more electronegative than phosphorus, so that the phosphorus–chlorine bond has the polarity shown in Figure 79. The positive end of the water molecule is attracted by the chlorine atom, and hydrolysis proceeds as shown.

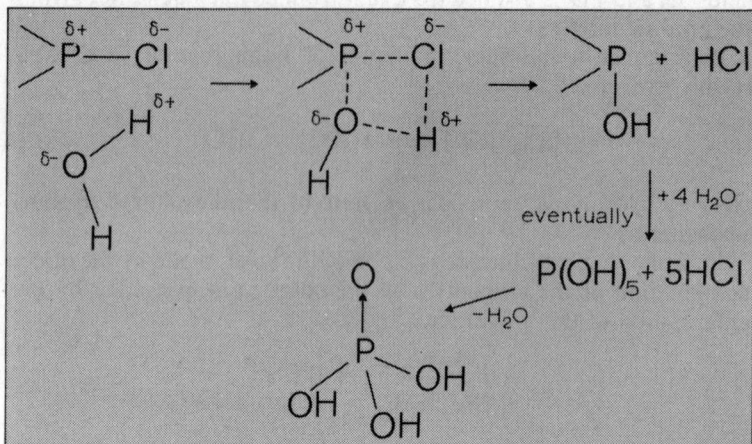

Fig. 79 Hydrolysis of phosphorus pentachloride

The intermediate structure $P(OH)_5$ loses a molecule of water to form H_3PO_4.

A similar series of stages can be worked out for the hydration of phosphorus trichloride.

In the case of nitrogen trichloride, the electronegativities are reversed, nitrogen is slightly more electronegative than the chlorine and the hydrolysis takes place along the lines shown in Figure 80. The final products are ammonia and chloric (I) acid.

In general, the halogen atom is more electronegative than the central non-metallic atom, and almost all the non-metal chlorides and fluorides hydrolyse to give products similar to those given by the chlorides of phosphorus; for example,

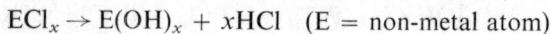

$$ECl_x \rightarrow E(OH)_x + xHCl \quad (E = \text{non-metal atom})$$

The $E(OH)_x$ may lose water in forming the stable oxyacid.

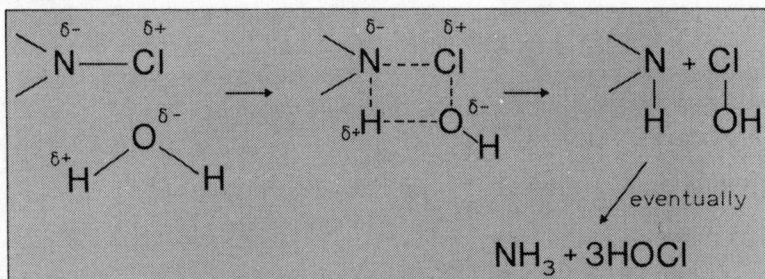

Fig. 80 Hydrolysis of nitrogen trichloride

Phosphorus trichloride oxide, $POCl_3$ This compound is made on a large scale by the oxidation of phosphorus trichloride with oxygen:

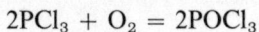

$$2PCl_3 + O_2 = 2POCl_3$$

It is used on a large scale for the manufacture of insecticides, solvents and plasticizers.

Oxides of Phosphorus

In the solid state, phosphorus is made up of tetrahedral P_4 units. On reacting with a controlled amount of oxygen, the phosphorus links break, and oxygen bridges are formed to produce P_4O_6.

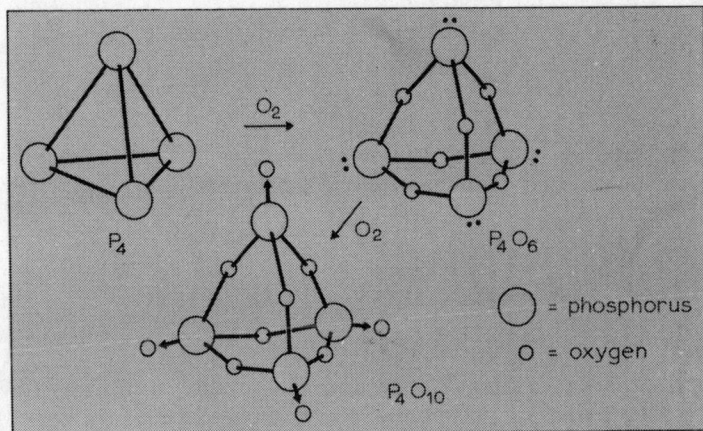

Fig. 81 Oxides of phosphorus

181

According to the electron pair repulsion theory, each phosphorus atom in the P_4O_6 structure has a lone pair of electrons directed as shown in Figure 81. Thus, on reaction with excess oxygen, P_4O_{10} is formed.

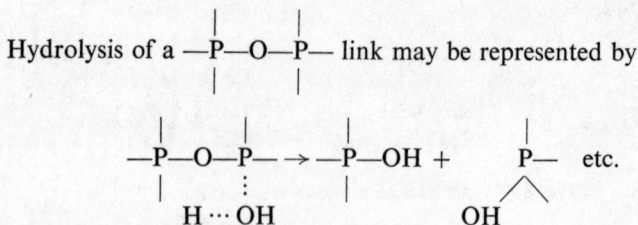

Hydrolysis of a $-\overset{|}{\underset{|}{P}}-O-\overset{|}{\underset{|}{P}}-$ link may be represented by

$$-\overset{|}{\underset{|}{P}}-O-\overset{|}{\underset{|}{P}}-\ \underset{H\cdots OH}{\vdots}\ \rightarrow\ -\overset{|}{\underset{|}{P}}-OH\ +\ \overset{|}{\underset{OH}{P}}-\ \text{etc.}$$

Thus, P_4O_{10} produces $P(OH)_5$. This is an unstable intermediate structure, which loses a molecule of water to form phosphoric (V) acid,

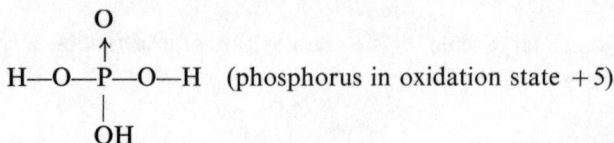

$$H-O-\overset{\overset{O}{\uparrow}}{\underset{\underset{OH}{|}}{P}}-O-H\quad\text{(phosphorus in oxidation state }+5)$$

P_4O_6 forms $P(OH)_3$ (phosphorous acid) on hydrolysis, which is a tetrahedral molecule

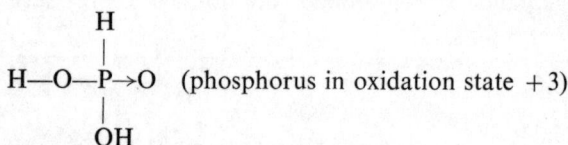

$$H-O-\overset{\overset{H}{|}}{\underset{\underset{OH}{|}}{P}}\!\rightarrow\!O\quad\text{(phosphorus in oxidation state }+3)$$

In an ionizing solvent, the oxygen–hydrogen bond undergoes heterolytic fission (i.e., the more electronegative oxygen atom retains both electrons).

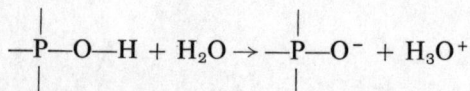

$$-\overset{|}{\underset{|}{P}}-O-H\ +\ H_2O\ \rightarrow\ -\overset{|}{\underset{|}{P}}-O^-\ +\ H_3O^+$$

The phosphorus–hydrogen bond does not rupture in this way (the electronegativities of phosphorus and hydrogen are almost equal). Thus phosphoric (V) acid has a basicity of three, but phosphorous acid is dibasic.

It was noted, from the oxidation state chart on page 177, that oxyacids of phosphorus in which the phosphorus has an oxidation state of less than $+5$ are reducing. Phosphorous acid has, therefore, reducing properties:

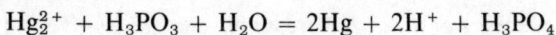

$$Hg_2^{2+} + H_3PO_3 + H_2O = 2Hg + 2H^+ + H_3PO_4$$

(In these reactions, the P—H bond is oxidized to P—O—H.)
Dehydration of phosphoric (V) acid leads to the formation of *condensed* acids (see p. 163):

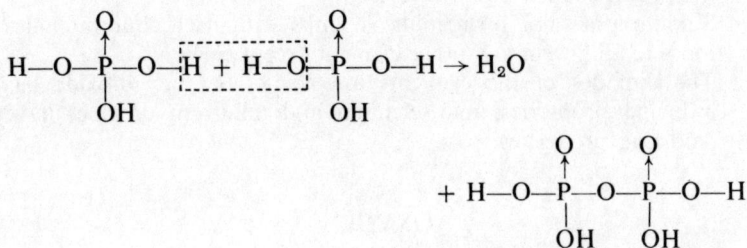

The resulting acid has a basicity of four, but it is not a reducing agent, as phosphorus is in oxidation state $+5$. It may be prepared by gently heating phosphoric (V) acid:

$$2H_3PO_4 = H_2O + H_4P_2O_7$$

Trioxo phosphoric acid is made by a more drastic dehydration of phosphoric (V) acid, using phosphorus trichloride oxide:

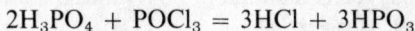

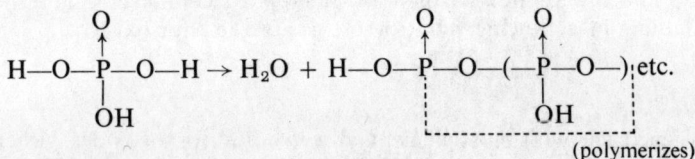

$$2H_3PO_4 + POCl_3 = 3HCl + 3HPO_3$$

The acid and the salts are polymerized, for example $Na_6(PO_3)_6$.

Alkali and alkaline earth phosphates are widely used in the manufacture of detergents.

GROUP VIB: OXYGEN, SULPHUR, SELENIUM AND TELLURIUM

General Properties

1 Early members of the group are highly electronegative and reactive. They can form anions carrying a double negative charge.
2 All the elements can show a covalency of two. This is the maximum with oxygen, but the remainder of the elements have a covalency maximum of six.
3 Oxygen can form multiple bonds with itself, but does not catenate (i.e., form chains) significantly.
4 Sulphur does not form multiple links with itself, but catenates more readily than any other element except carbon.
5 The dioxides of the elements are acidic. Sulphur dioxide has reducing properties, but selenium and tellurium dioxides have oxidizing properties.

OXYGEN

Oxygen is prepared in the laboratory by the action of acidified potassium manganate (VII) on hydrogen peroxide:

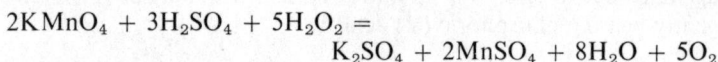

$$2KMnO_4 + 3H_2SO_4 + 5H_2O_2 =$$
$$K_2SO_4 + 2MnSO_4 + 8H_2O + 5O_2$$

It may also be prepared by the action of sodium peroxide on water, by the action of a cobalt catalyst on chlorate (I) solutions, or by the electrolysis of dilute acid or dilute alkali solutions.

Industrially, oxygen is manufactured by the fractionation of liquid air.

Oxygen is a colourless, odourless gas, which is sparingly soluble in water and slightly heavier than air. It supports combustion, and the rekindling of a glowing splint by the gas is a test for oxygen.

Oxides

Oxygen reacts with most elements directly, and forms oxides, either directly or indirectly, with almost every known element. In general, the methods available for preparing oxides are:

(a) Direct combination—metals sometimes oxidize superficially.
(b) By dissolving a metal in nitric acid and heating the nitrate.

184

(c) By heating a metal hydroxide or carbonate.
(d) By the action of a strong oxidizing agent on the element, for example, tin and nitric acid.
(e) By the dehydration of an acid.

Oxides may be divided into the following classes:

1 *Basic oxides*. These are metallic oxides (with the metal in a low oxidation state) which react with acids to give a salt and water only:

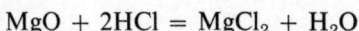

$$MgO + 2HCl = MgCl_2 + H_2O$$

2 *Acidic oxides*. These are oxides of non-metals (and some metals such as the transition metals in high oxidation states), and they are usually soluble in water, producing acids. Hence, acidic oxides are often referred to as acid anhydrides. They react with alkalies to form a salt and water only:

$$2NaOH + SO_2 = Na_2SO_3 + H_2O$$

3 *Amphoteric oxides*. These oxides, which show both basic and acidic character, are usually those of the less electropositive metals:

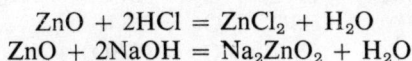

$$ZnO + 2HCl = ZnCl_2 + H_2O$$
$$ZnO + 2NaOH = Na_2ZnO_2 + H_2O$$

4 *Neutral oxides*. These are oxides which show neither acidic nor basic properties. Examples are carbon monoxide and water.

5 *Peroxides*. These oxides liberate hydrogen peroxide when treated with water or dilute acid:

$$BaO_2 + 2H^+ = Ba^{2+} + H_2O_2$$

(In addition, the more reactive alkali metals form *superoxides* containing the ion O_2^-. They are formed by heating the metal directly in oxygen, e.g., $K + O_2 = KO_2$.)

Uses (a) The major use of oxygen is as 'tonnage' oxygen in steelmaking.
(b) In the manufacture of ethene oxide and other chemicals from petroleum.
(c) As a rocket fuel.
(d) In oxy-acetylene welding and cutting.

Ozone (Trioxygen)

Ozone is prepared by passing a silent electric discharge through oxygen in an apparatus known as an ozonizer (Figure 82).

$$3O_2 \rightleftharpoons 2O_3$$

Ozone is a pale blue gas, which is moderately soluble in water. It is used in sterilizing water, and for purifying air. Mercury, in the presence of ozone, loses its meniscus, and adheres to glass.

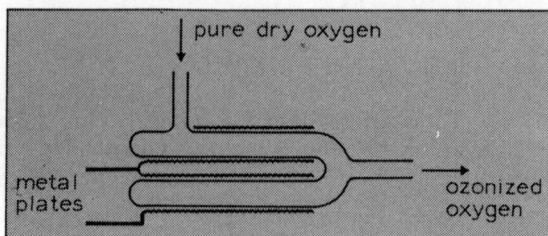

Fig. 82 Ozonizer

Ozone is decomposed catalytically by alkali solutions, some metals and by some metal oxides. Ozone is a powerful oxidizing agent; for example:

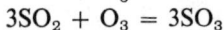

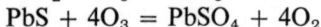

$$3SnCl_2 + 6HCl + O_3 = 3SnCl_4 + 3H_2O$$
$$3SO_2 + O_3 = 3SO_3$$
$$PbS + 4O_3 = PbSO_4 + 4O_2$$

The gas oxidizes many organic compounds, sometimes explosively.
Ozone adds on to alkene-type double bonds in organic compounds:

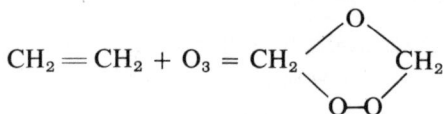

$$CH_2 = CH_2 + O_3 = CH_2 \underset{O-O}{\overset{O}{\diagup \diagdown}} CH_2$$

The formation of these ozonides can be used to determine the number of double bonds present in a molecule.

Ozone may be determined quantitatively by the liberation of iodine from neutral potassium iodide solution:

$$O_3 + 2KI + H_2O = I_2 + 2KOH + O_2$$

The formula of ozone is confirmed by *Soret's experiment*. In this experiment, the percentage of ozone in a sample of ozonized air is found by measuring the loss in volume when the sample is shaken with a liquid alkene. Another sample of the same ozonized air is heated to decompose the ozone and form oxygen.

186

If the formula of ozone is O_n, then on heating,

$$O_n \rightarrow \frac{n}{2} O_2$$

and so, 1 cm³ of ozone produces $n/2$ cm³ of oxygen.

By experiment, it is found that $1\frac{1}{2}$ cm³ of oxygen are produced from 1 cm³ of ozone. Thus,

$$n/2 = 3/2 \text{ and } n = 3$$

Hydrogen Peroxide

A solution of hydrogen peroxide is produced in the laboratory by the acidification of a peroxide:

$$BaO_2 + H_2SO_4 = BaSO_4 + H_2O_2$$

Industrially, hydrogen peroxide is manufactured:

(a) By the electrolysis of ammonium sulphate solution, acidified with sulphuric acid, using inert electrodes. At the anode:

$$2SO_4^{2-} \rightarrow S_2O_8^{2-} + 2e^-$$

the electrons going to the anode. When the electrolyte solution is heated:

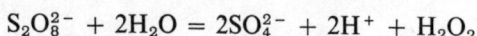

$$S_2O_8^{2-} + 2H_2O = 2SO_4^{2-} + 2H^+ + H_2O_2$$

(b) By the autoxidation of an anthraquinone:

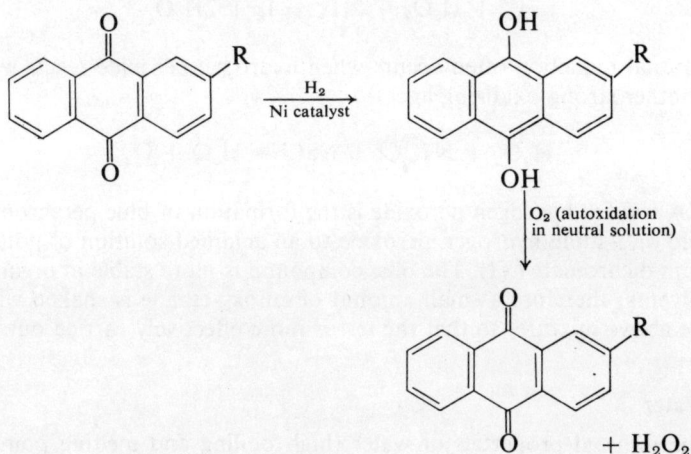

187

In both methods, the hydrogen peroxide produced is concentrated by vacuum distillation. Because this is a dangerous operation in concentrations greater than 80 per cent, the final concentration is achieved by crystallization.

Hydrogen peroxide is usually available in solutions of a given *volume strength*. This shows the volume of oxygen produced (at s.t.p.) by the decomposition of one volume of hydrogen peroxide according to the equation:

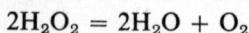

$$2H_2O_2 = 2H_2O + O_2$$

The major use of hydrogen peroxide is in bleaching; it is also used in the manufacture of organic peroxides (which function as 'initiators' in polymerization reactions) and as a rocket fuel.

Properties

Hydrogen peroxide is a colourless, odourless syrupy liquid (melting point $-1 \cdot 7°C$). It is unstable, and is decomposed by catalysts such as manganese dioxide or finely divided silver:

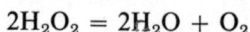

$$2H_2O_2 = 2H_2O + O_2$$

Hydrogen peroxide is a strong oxidizing agent.

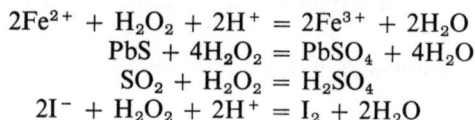

$$2Fe^{2+} + H_2O_2 + 2H^+ = 2Fe^{3+} + 2H_2O$$
$$PbS + 4H_2O_2 = PbSO_4 + 4H_2O$$
$$SO_2 + H_2O_2 = H_2SO_4$$
$$2I^- + H_2O_2 + 2H^+ = I_2 + 2H_2O$$

Mutual reduction often occurs when hydrogen peroxide reacts with another strong oxidizing agent:

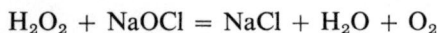

$$H_2O_2 + NaOCl = NaCl + H_2O + O_2$$

A test for hydrogen peroxide is the formation of blue perchromic acid on adding hydrogen peroxide to an acidified solution of potassium dichromate (VI). The blue compound is more stable in organic solvents; therefore a small amount of ethoxy-ethane is shaken with the above mixture, so that the test is more effectively carried out.

Water

The unusual properties of water (high boiling and melting points,

188

high latent heat of fusion and vaporization, surface tension, etc.) are explained in terms of hydrogen bonding.

Water undergoes self-ionization:

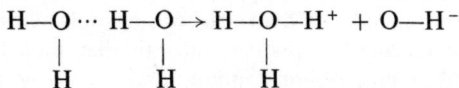

$$H—O \cdots H—O \rightarrow H—O—H^+ + O—H^-$$
$$\underset{H}{\vert} \quad \underset{H}{\vert} \quad \underset{H}{\vert}$$

The degree of ionization is slight, and a hydroxide ion and a hydrated proton are produced.

Hard water The presence of temporary hardness (which may be removed by boiling) has already been mentioned on p. 159. Permanent hardness, which is not removed by boiling, is due to the presence of salts such as calcium and magnesium sulphates in water. The degree of hardness in water is usually estimated by titration with EDTA which has been standardized by titrating against a calcium solution of known concentration.

Hardness in water is removed by:

(a) Addition of sodium carbonate:

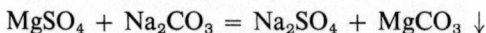

$$MgSO_4 + Na_2CO_3 = Na_2SO_4 + MgCO_3 \downarrow$$

(b) Ion exchange.

A cation exchange resin can accept metal ions and liberate an equivalent number of hydrogen ions (e.g., Mg^{2+} accepted, $2H^+$ liberated). An anion exchange resin operates similarly—for example, Cl^- accepted, OH^- released. By passing water over a cation exchange resin, soft water may be produced. Using a *mixed bed* resin, containing both cation and anion resins, de-ionized water is produced. This water has a very low electrical conductivity, although it may contain non-ionic impurities. The ion exchange resins may be regenerated by soaking in acid (for a cation resin) or alkali (for anion resins), for some time.

SULPHUR

Sulphur is found in the free state in underground deposits, especially in the USA. It is obtained by the *Frasch method*, in which three concentric pipes are sunk into the ground to reach the deposit. Superheated steam is forced down the outer pipe and compressed air

189

is blown down the central shaft. The molten sulphur at the base of the pipes forms a froth with the compressed air and rises to the surface. This sulphur is over 99 per cent pure.

An alternative source of sulphur is found in petroleum crudes. When these are refined, the hydrocarbons of high relative molecular mass are split (*catalytic cracking*) into smaller molecules. At this stage, all sulphur containing compounds have to be removed (by dissolution in an organic solvent) and in this way, quantities of sulphur are made available.

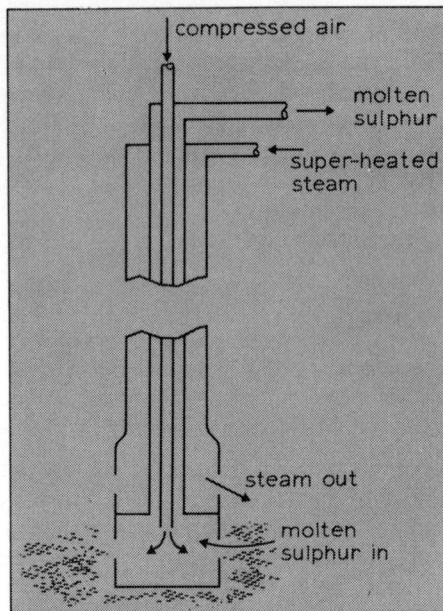

Fig. 83 Frasch process

Allotropes of Sulphur

1 *Rhombic* (or α) sulphur crystallizes out from a solution of sulphur in carbon disulphide or methylbenzene. The structural units are S_8 rings packed as shown in Figure 84. This form is stable up to 97°C.
2 *Monoclinic* or *prismatic* (or β) sulphur is made by allowing a crust to form on the surface of a cooling mass of molten sulphur. Two holes are pierced in the crust, and the remaining molten sulphur is drained away. Needle-shaped crystals of monoclinic sulphur are found on the underside of the crust. The structural

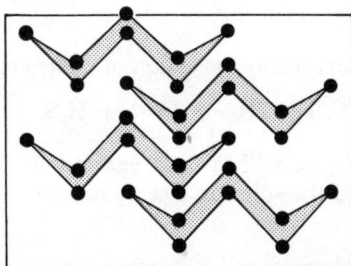

Fig. 84 Structure of rhombic (or α) sulphur

unit is again an S_8 ring, but the packing of these rings takes a more elongated form. Monoclinic sulphur is stable between 97°C and the melting point, 119°C.

On melting sulphur, a golden, free flowing liquid is first formed. This corresponds to the free movement of complete S_8 rings in the liquid. On further heating, the viscosity increases, and at 190°C the liquid is almost black. This corresponds to the breaking of the S_8 rings to form long chains which become completely entangled. (Sulphur has a tendency to catenate.) As the temperature increases still further, the viscosity decreases because the chains break up and, probably, form S_8 rings once more. The vapour of sulphur just above the boiling point, 444·4°C, has been shown to consist of S_8, S_6, S_4 and S_2 units.

3 When the molten sulphur at the very viscous stage is quickly cooled by pouring into water, the rubbery *plastic sulphur* is formed. This again is thought to contain long chains of sulphur atoms, which slowly revert back to S_8 units.

4 *Colloidal* sulphur is made by the acidification of sodium thiosulphate (VI) solution with dilute acid.

Reactions of Sulphur

1 Burns on heating in air:

$$S + O_2 = SO_2$$

2 No reaction with water and dilute acids.

3 Concentrated nitric acid oxidizes sulphur:

$$S + 6HNO_3 = H_2SO_4 + 6NO_2 + 2H_2O$$

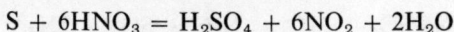

4 Many metals react directly with sulphur on heating:

$$Fe + S = FeS$$

191

Hydrogen Sulphide

The gas is usually prepared by the action of dilute acid on a sulphide:

$$FeS + 2H^+ = Fe^{2+} + H_2S$$

It is a poisonous gas, with a 'bad egg' odour. It is moderately soluble in water, and the solution is a weak acid; therefore it reacts with alkali to form:

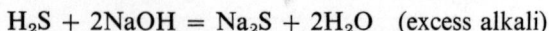

$$H_2S + 2NaOH = Na_2S + 2H_2O \quad \text{(excess alkali)}$$

then

$$H_2S + Na_2S = 2NaHS$$

Hydrogen sulphide burns in air with a pale blue flame:

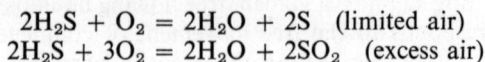

$$2H_2S + O_2 = 2H_2O + 2S \quad \text{(limited air)}$$
$$2H_2S + 3O_2 = 2H_2O + 2SO_2 \quad \text{(excess air)}$$

Hydrogen sulphide is a reducing agent:

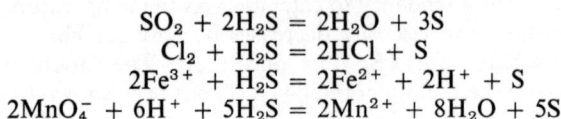

$$SO_2 + 2H_2S = 2H_2O + 3S$$
$$Cl_2 + H_2S = 2HCl + S$$
$$2Fe^{3+} + H_2S = 2Fe^{2+} + 2H^+ + S$$
$$2MnO_4^- + 6H^+ + 5H_2S = 2Mn^{2+} + 8H_2O + 5S$$

The gas precipitates metal sulphides from solution:

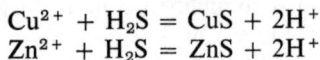

$$Cu^{2+} + H_2S = CuS + 2H^+$$
$$Zn^{2+} + H_2S = ZnS + 2H^+$$

Precipitation of these sulphides takes place in acid or alkaline conditions, depending on the solubility product of the individual metal sulphide.

Sulphur Dioxide

The gas is prepared by any of the following methods:

(a) The acidification of a sulphite, or hydrogen sulphite:

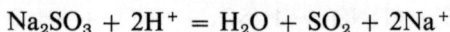

$$Na_2SO_3 + 2H^+ = H_2O + SO_2 + 2Na^+$$

(b) Burning sulphur in air:

$$S + O_2 = SO_2$$

(c) The reaction between hot, concentrated sulphuric acid and a metal such as copper or zinc:

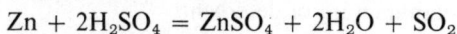

$$Zn + 2H_2SO_4 = ZnSO_4 + 2H_2O + SO_2$$

(some hydrogen may be produced also).

Sulphur dioxide is a colourless gas with a choking smell. It is heavier than air, and is quite soluble in water, the solution containing sulphurous acid. Sulphur dioxide does not burn, nor support combustion, although magnesium will burn in it:

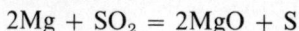

$$2Mg + SO_2 = 2MgO + S$$

On passing sulphur dioxide through an alkaline solution:

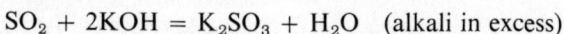

$$SO_2 + 2KOH = K_2SO_3 + H_2O \quad \text{(alkali in excess)}$$

then,

$$SO_2 + K_2SO_3 + H_2O = 2KHSO_3$$

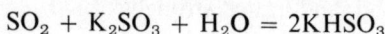

Sulphur dioxide turns calcium hydroxide solution milky by the formation of calcium sulphite. However, the precipitate clears quickly because of the ready solubility of sulphur dioxide and the rapid formation of calcium hydrogen sulphite. For this reason, sulphur dioxide is not often confused with carbon dioxide during the application of the 'lime water' test.

Sulphur dioxide is a reducing agent; potassium manganate (VII) solution is decolourized, and dichromate is turned green:

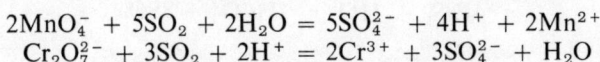

$$2MnO_4^- + 5SO_2 + 2H_2O = 5SO_4^{2-} + 4H^+ + 2Mn^{2+}$$
$$Cr_2O_7^{2-} + 3SO_2 + 2H^+ = 2Cr^{3+} + 3SO_4^{2-} + H_2O$$

Sulphurous acid and acidified sulphite solutions also reduce:

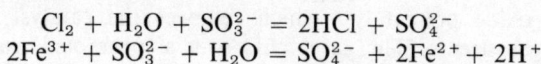

$$Cl_2 + H_2O + SO_3^{2-} = 2HCl + SO_4^{2-}$$
$$2Fe^{3+} + SO_3^{2-} + H_2O = SO_4^{2-} + 2Fe^{2+} + 2H^+$$

Sulphur dioxide is used in the manufacture of sulphuric acid, as a solvent (in liquefied form), in petroleum refining and as a bleach and preservative.

Sulphur Trioxide

Sulphur trioxide may be prepared by the reaction of sulphur dioxide with oxygen in the presence of a catalyst, or by the dehydration of sulphuric acid by phosphorus pentachloride:

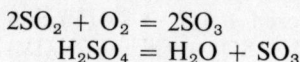

$$2SO_2 + O_2 = 2SO_3$$
$$H_2SO_4 = H_2O + SO_3$$

Sulphur trioxide is a white, volatile solid at s.t.p. It is available commercially in a specially stabilized liquid form, and this is widely used for the manufacture of detergents.

Sulphuric Acid

Two processes are used for the manufacture of this very important chemical:

1 the contact process,
2 the lead chamber process.

Both methods rely on the conversion of sulphur dioxide to the trioxide.

The main source of sulphur is anhydrite, $CaSO_4$. This is mixed with coke and sand, and heated strongly:

$$CaSO_4 + 2C = CaS + 2CO_2$$

Then,

$$CaS + 3CaSO_4 = 4CaO + 4SO_2$$
$$CaO + SiO_2 = CaSiO_3$$

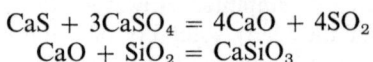

(a by-product used in cement and plaster manufacture).

Sulphur dioxide is also produced by burning sulphur in air, or by heating pyrites or other similar sulphide ores:

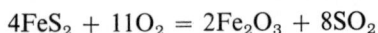

$$4FeS_2 + 11O_2 = 2Fe_2O_3 + 8SO_2$$

The contact process Dry, purified and dust-free sulphur dioxide is mixed with an excess of oxygen, and the mixture is passed over an iron (III) oxide pre-catalyst which removes the catalyst poisons from the gas stream. The gases pass over a 'promoted' vanadium catalyst which incorporates alkali metal sulphates. The catalyst is made up using an inert support material such as silica gel and the catalyst granules are activated by exposing them to sulphur dioxide and air at 400–500°C. This forms a thin liquid film of alkali metal sulphur compounds which complex the vanadium and hold it in solution. A reduction in temperature below a critical level causes a sharp decrease in catalytic activity.

The reactions, which take place in the liquid phase at the surface of the catalyst granules may be formulated as:

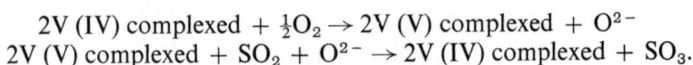

$$2V \text{ (IV) complexed} + \tfrac{1}{2}O_2 \rightarrow 2V \text{ (V) complexed} + O^{2-}$$
$$2V \text{ (V) complexed} + SO_2 + O^{2-} \rightarrow 2V \text{ (IV) complexed} + SO_3.$$

194

Heat is evolved in the overall reaction, so the gases are cooled between successive beds of catalysts.

After cooling, the sulphur trioxide is absorbed in concentrated sulphuric acid, producing oleum or fuming sulphuric acid. This is demanded by certain industrial processes, but most of the oleum is diluted to give the concentrated acid.

The lead chamber process   Nitrogen oxide is used as a catalyst, the reactions being:

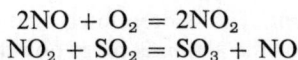

$$2NO + O_2 = 2NO_2$$
$$NO_2 + SO_2 = SO_3 + NO$$

Sulphur dioxide is mixed with excess air, and the mixture is passed into the first of a series of lead-lined towers. The mixture of oxides of nitrogen which serves as a catalyst is admitted, and the above reactions take place. In a subsequent tower, a water spray is used to dissolve the sulphur trioxide:

$$SO_3 + H_2O = H_2SO_4$$

The remaining gases are recycled until conversion is complete. The process is worked to give an acid concentration of about 75 per cent. About one-third of the plants manufacturing sulphuric acid still use the lead chamber process, although the method is gradually being displaced by the contact process.

Uses   (a) For the conversion of calcium phosphate to the 'super-phosphate' which is used as a fertilizer.
(b) For the production of ammonium sulphate, which is also used as a fertilizer.
(c) As a sulphonating agent, in the production of surface-active agents.
(d) As an acid bath in the manufacture of man-made fibres.
(e) In the dyestuffs industry.
(f) In the manufacture of pigments.
(g) In the manufacture of explosives.

Reactions of Sulphuric Acid

These may be grouped under four headings:

1 *As an acid*. It reacts as a dibasic acid, forming both normal and acid salts. For example:

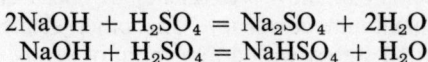

$$2NaOH + H_2SO_4 = Na_2SO_4 + 2H_2O$$
$$NaOH + H_2SO_4 = NaHSO_4 + H_2O$$

Most sulphates are soluble. Notable exceptions are calcium, strontium, barium and lead sulphates.

2 *As a dehydrating agent.* Concentrated sulphuric acid is a powerful dehydrating agent and is used in drying gases. It reacts vigorously with carbohydrates, producing charring, and leaving a mass of carbon. For example, with glucose:

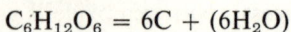

$$C_6H_{12}O_6 = 6C + (6H_2O)$$

3 *As an oxidizing agent.* The concentrated acid is a mild oxidizing agent when cold, but the action is more powerful on heating. For example:

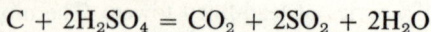

$$C + 2H_2SO_4 = CO_2 + 2SO_2 + 2H_2O$$

4 *As a sulphonating agent.* On refluxing benzene with 100 per cent sulphuric acid, benzene sulphonic acid is formed:

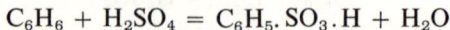

$$C_6H_6 + H_2SO_4 = C_6H_5.SO_3.H + H_2O$$

Sulphamic Acid, $(NH_2)(OH)SO_2$

This acid is made from urea, $NH_2.CO.NH_2$, by the following reaction:

$$NH_2.CO.NH_2 + SO_3 + H_2SO_4 = 2NH_2.SO_2.OH + CO_2$$

Sulphamic acid is the only strong acid available in crystalline form. It is used for cleaning and descaling many items of chemical plant, and as a sulphonating agent, especially in the production of artificial sweeteners.

GROUP VIIB: THE HALOGENS (FLUORINE, CHLORINE BROMINE AND IODINE)

General Properties

1 Highly electronegative elements, all capable of forming an anion bearing a single negative charge.
2 Their electronegativity decreases with increasing atomic size, fluorine being the most electronegative.
3 All the elements can show a covalency of one. For fluorine, this is also its maximum covalency. The other elements can show higher oxidation states, with a maximum of seven.

4 All the elements exist as diatomic molecules.
5 There is a steady increase in melting points and boiling points with increasing atomic number.
6 The elements react progressively less vigorously with hydrogen.

FLUORINE

Fluorine, in common with the first element in any group of the Periodic Table, shows significant differences in properties compared with the other members of the group.

1 Fluorine is extremely reactive. This is due to

(a) The low fluorine–fluorine bond strength in the fluorine molecule.
(b) The high bond strength of the bond formed between fluorine and most other elements.
(c) The strong electron-attracting power of a fluorine atom.
(d) The small size of the fluorine atom, and the fluoride ion. As a result of factors (b) and (c), fluorine is regarded as capable of forcing another element to display its maximum covalency.

2 The oxidation state of fluorine is always -1, and fluorine always shows a covalency of one.
3 Fluorine does not form oxyacids.
4 Fluorinated hydrocarbons are resistant to further reaction, because of the high carbon–fluorine bond strength.
5 The solubility of fluorides is often quite different from the solubility of the other halides of the same element, for example, silver fluoride is soluble, while the other silver halides are not. This is an observation that may be linked with the higher hydration energy of the fluoride ion.

Preparation and Properties of Fluorine

The source of fluorine is calcium fluoride, CaF_2 (fluorspar), and from this compound, hydrogen fluoride is prepared:

$$CaF_2 + H_2SO_4 = CaSO_4 + 2HF$$

Electrolysis of potassium hydrogen fluoride, KHF_2, in anhydrous hydrogen fluoride liberates fluorine at the anode:

$$2F^- = 2e^- + F_2$$

197

The electrolytic cell is made of steel, fitted with a copper-impregnated graphite anode. Hydrogen is evolved at the cathode.

Fluorine is an extremely reactive gas. It explodes with hydrogen under all conditions (provided a trace of moisture is present):

$$H_2 + F_2 = 2HF$$

Fluorine reacts with most elements directly.

Hydrogen fluoride is prepared by the action of concentrated sulphuric acid on calcium fluoride. It is a gas (b.p. $19 \cdot 5°C$) which is very soluble in water, but the solution is a weak acid. This is due to the high fluorine–hydrogen bond strength, and the energy released on the hydration of the hydrogen and fluoride ions produced by the heterolytic fission of the bond is insufficient to promote extensive bond rupture.

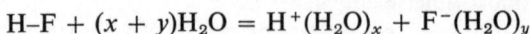

$$H\text{–}F + (x + y)H_2O = H^+(H_2O)_x + F^-(H_2O)_y$$

Hydrofluoric acid is poisonous and corrosive. It attacks glass:

$$SiO_2 + 4HF = SiF_4 + 2H_2O$$

Also,

$$2HF + SiF_4 = H_2SiF_6$$

Hydrogen fluoride is used for the production of fluorides, for etching glass and metals, and in the extraction of uranium from its oxide ore.

CHLORINE

The chief source of chlorine is sodium chloride.

The laboratory preparation of chlorine involves the oxidation of concentrated hydrochloric acid by manganese dioxide or potassium manganate (VII):

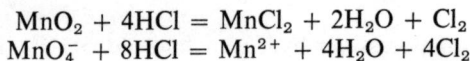

$$MnO_2 + 4HCl = MnCl_2 + 2H_2O + Cl_2$$
$$MnO_4^- + 8HCl = Mn^{2+} + 4H_2O + 4Cl_2$$

Chlorine is available as a by-product in the electrolytic manufacture of sodium hydroxide (see p. 92).

198

Properties

Chlorine is a greenish-yellow gas with a choking smell. It is toxic, and much heavier than air. It dissolves in water (being moderately soluble) to form a mixture of hydrochloric and chloric (I) acids. Hence, on bubbling through alkali the following reaction takes place:

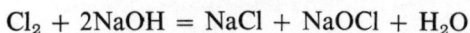

$$Cl_2 + 2NaOH = NaCl + NaOCl + H_2O$$

If the solution is hot, and the concentration of sodium hydroxide is high, sodium chlorate (V) is formed on the passage of an excess of chlorine into the solution:

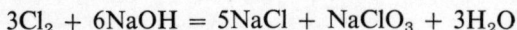

$$3Cl_2 + 6NaOH = 5NaCl + NaClO_3 + 3H_2O$$

Chlorine reacts directly with most metals and with a number of non-metals. The reaction with hydrogen is explosive in sunlight, and it is regarded as a chain reaction (see p. 146).

There are many important reactions of chlorine in organic chemistry, especially in connection with the production of plastics, solvents and insecticides.

Chlorine is used for the production of hydrochloric acid, in bleaching and sterilizing, and in the production of organic chloro compounds.

Hydrogen Chloride

This compound is made in the laboratory by the action of concentrated sulphuric acid on a chloride:

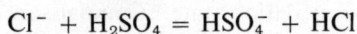

$$Cl^- + H_2SO_4 = HSO_4^- + HCl$$

On a large scale, hydrogen chloride is produced by burning chlorine in hydrogen. A considerable quantity of hydrogen chloride is available as a by-product in the chlorination of organic compounds.

Pure, dry hydrogen chloride is a gas which shows no acidic properties. It is very soluble in water, and the solution is a strong acid (hydrochloric acid). The acid strengths of solutions of hydrogen bromide and hydrogen iodide increase progressively, due to the weaker hydrogen–halogen bond strength:

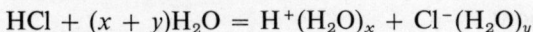

$$HCl + (x + y)H_2O = H^+(H_2O)_x + Cl^-(H_2O)_y$$

Hydrochloric acid forms a constant boiling mixture (b.p. 110°C) containing 20 per cent of the acid by weight. It is the parent acid of the chlorides, all of which are soluble except those of lead, mercury (I) and silver.

Oxyacids of Chlorine

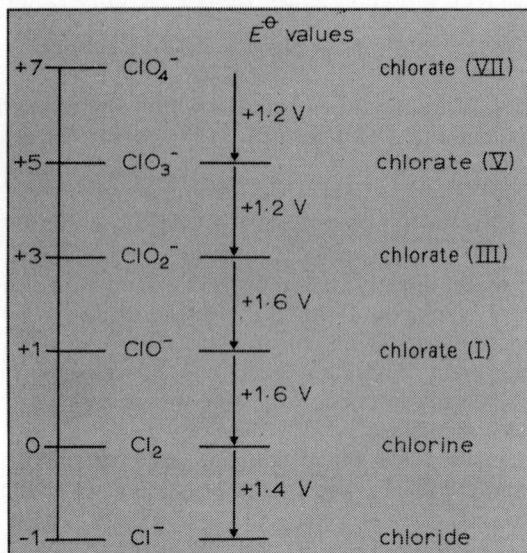

Fig. 85 Oxidation state chart for chlorine

From the oxidation state chart, all the positive oxidation states of chlorine are oxidizing agents, and the preferred tendency for chlorine is to achieve a low oxidation state. (Comparison of this chart with that for phosphorus explains the violent reaction between chlorate (V) and phosphorus.)

Chlorate (III) ion would be expected to undergo disproportionation.

Chlorate (I) compounds   These are made either by passing chlorine into cold dilute alkali, or by the electrolysis of cold dilute sodium chloride, under conditions such that the chlorine liberated by electrolysis mixes thoroughly with the electrolyte.

Chlorate (I) yields chlorine on acidification in the presence of chloride ion:

$$ClO^- + Cl^- + 2H^+ = Cl_2 + H_2O$$

Oxygen is evolved on warming chlorate (I) with a cobalt salt as a catalyst:

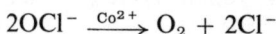

$$2OCl^- \xrightarrow{Co^{2+}} O_2 + 2Cl^-$$

Chlorate (I) displaces iodine from iodides and bromine from bromides in acid solution:

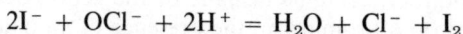

$$2I^- + OCl^- + 2H^+ = H_2O + Cl^- + I_2$$

Chloric (V) acid and chlorate (V) Chlorate (V) is made by the electrolysis of a saturated solution of sodium chloride, while a solution of chloric (V) acid may be prepared by the acidification of barium chlorate (V) with dilute sulphuric acid:

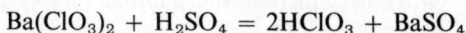

$$Ba(ClO_3)_2 + H_2SO_4 = 2HClO_3 + BaSO_4$$

Sodium chlorate (V) is used as a weedkiller, while the action of concentrated sulphuric acid (in the presence of an ethanedioate) on sodium chlorate (V) produces chlorine dioxide, ClO_2. Chlorine dioxide is a highly explosive gas; it is used in the preparation of chlorate (III) salts and as a bleach.

Chloric (VII) acid The electrolytic oxidation of sodium chlorate (V) solution produces sodium chlorate (VII). Acidification of sodium chlorate (VII) with concentrated sulphuric acid, followed by vacuum distillation, yields chloric (VII) acid:

$$2NaClO_4 + H_2SO_4 = Na_2SO_4 + 2HClO_4$$

This acid is the strongest acid known (i.e., the most completely ionized), and it is a very powerful oxidizing agent. It oxidizes organic materials with explosive violence. Chloric (VII) acid is also the most stable oxyacid of chlorine.

BROMINE

Bromine is found in some salt deposits, and it is present in sea water. It is extracted from bromide ores by displacement with chlorine:

$$2Br^- + Cl_2 = 2Cl^- + Br_2$$

In the extraction of bromine from sea water, the sea water is acidified with sulphuric acid, and chlorine gas is again used to liberate the bromine. The liberated bromine is either condensed, or reacted with sulphur dioxide:

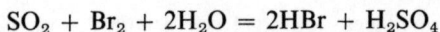

$$SO_2 + Br_2 + 2H_2O = 2HBr + H_2SO_4$$

In the laboratory, bromine is made by the action of concentrated sulphuric acid on a bromide, in the presence of an oxidizing agent:

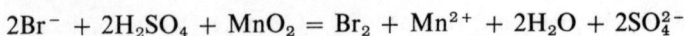

$$2Br^- + 2H_2SO_4 + MnO_2 = Br_2 + Mn^{2+} + 2H_2O + 2SO_4^{2-}$$

The main use of bromine is in the production of brominated petrol additives which eliminate the build-up of lead from anti-knock preparations. Other uses include the manufacture of photographic materials and pharmaceuticals.

Hydrogen Bromide

Hydrogen bromide is made by the hydrolysis of phosphorus tribromide. This is brought about by adding bromine, drop by drop, to a paste of red phosphorus and water:

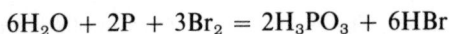

$$6H_2O + 2P + 3Br_2 = 2H_3PO_3 + 6HBr$$

Hydrogen and bromine react smoothly when heated to 200°C and in the presence of a catalyst:

$$H_2 + Br_2 = 2HBr$$

The reaction of bromine with sodium hydroxide produces sodium bromate (V) and sodium bromide (bromate (I) is not stable):

$$3Br_2 + 6NaOH = 5NaBr + NaBrO_3 + 3H_2O$$

IODINE

Most iodine is obtained from the Chilean deposits of sodium iodate (V). It is extracted by adding sodium hydrogen sulphite to the sodium iodate (V) solution:

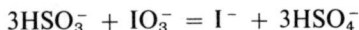

$$3HSO_3^- + IO_3^- = I^- + 3HSO_4^-$$

By using a calculated quantity of sodium hydrogen sulphite, only five-sixths of the iodate (V) is reduced. Then,

$$5I^- + IO_3^- + 6H^+ = 3I_2 + 3H_2O$$

The liberated iodine is filtered and purified by sublimation.

Iodine is sparingly soluble in water, but more soluble in potassium iodide solution, because of the formation of the tri-iodide complex ion:

$$KI + I_2 = KI_3$$

Iodine is more soluble in organic solvents

Iodine and hydrogen react reversibly on heating:

$$H_2 + I_2 \rightleftharpoons 2HI$$

Hydrogen iodide is made by the hydrolysis of phosphorus tri-iodide which is brought about by adding water, drop by drop, to a mixture of red phosphorus and iodine:

$$PI_3 + 3H_2O = 3HI + H_3PO_3$$

Iodine can act as a reducing agent:

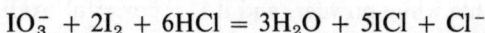

$$IO_3^- + 2I_2 + 6HCl = 3H_2O + 5ICl + Cl^-$$

This reaction takes place in concentrated hydrochloric acid solution, and iodine monochloride (iodine in the $+1$ oxidation state) is formed.

Iodine can also act as an oxidizing agent:

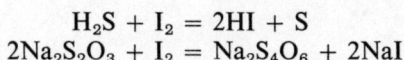

$$H_2S + I_2 = 2HI + S$$
$$2Na_2S_2O_3 + I_2 = Na_2S_4O_6 + 2NaI$$

Iodine is liberated when iodate reacts with iodide in dilute acid solution:

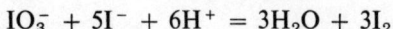

$$IO_3^- + 5I^- + 6H^+ = 3H_2O + 3I_2$$

The uses of iodine are widespread, ranging from medical and photographic to the manufacture of dyes.

Iodates

The reaction of hot concentrated potassium hydroxide solution with iodine produces potassium iodate (V):

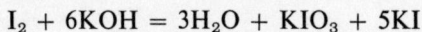

$$I_2 + 6KOH = 3H_2O + KIO_3 + 5KI$$

The potassium iodate (V) separates out on cooling, and potassium iodide is recovered from the filtrate.

Potassium iodate (V) can be obtained in a very pure state, and is used to liberate a known amount of iodine by the reaction with iodide ion in dilute acid, according to the equation:

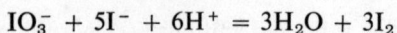

$$IO_3^- + 5I^- + 6H^+ = 3H_2O + 3I_2$$

This is especially useful in volumetric analysis, for the standardization of thiosulphate (VI) solutions.

Iodic (V) Acid, HIO_3

This is produced by the action of fuming nitric acid on iodine, or by the oxidation (using chlorine) of a suspension of iodine in water:

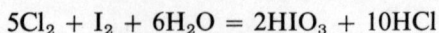

$$5Cl_2 + I_2 + 6H_2O = 2HIO_3 + 10HCl$$

The dehydration of iodic (V) acid by heating produces iodine pentoxide:

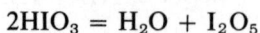

$$2HIO_3 = H_2O + I_2O_5$$

This is a stable white powder, and it is a powerful oxidizing agent.

Appendix

SUCCESSIVE IONIZATION ENERGIES OF SELECTED ELEMENTS
(Values expressed in MJ/mol)

Element	1	2	3	4	5	6	7	8	9	10
H	1·31									
He	2·37	5·24								
Li	0·53	7·28	11·79							
Be	0·90	1·75	14·81	20·9						
B	0·80	2·42	3·65	24·9	32·9					
C	1·08	2·35	4·61	6·21	37·7	47·2				
N	1·40	3·85	4·57	7·46	9·42	53·1	64·3			
O	1·32	3·38	4·29	7·45	10·97	13·28	70·1	83·8		
F	1·67	3·37	6·07	8·37	10·97	15·12	17·81	91·8	106·2	
Ne	2·08	3·94	6·11	9·34	12·13	15·21	19·93	23·0	115·0	131·1
Na	0·49	4·55	6·93	9·53	13·28	16·56	20·0	25·4	28·9	141·0
Mg	0·73	1·44	7·70	10·49	13·57	17·91	21·6	25·6	31·6	35·3
Al	0·58	1·73	2·69	11·55	14·83	18·29	23·2	27·3	31·8	38·4
Si	0·77	1·57	3·22	4·33	16·08	19·74	23·7	29·2	33·8	38·6
P	1·01	1·89	2·89	4·91	6·36	19·45	25·3	29·7	35·8	40·8
S	1·00	2·25	3·37	4·52	6·98	8·47	27·1	31·7	36·5	43·0
Cl	1·25	2·31	3·85	5·15	6·55	9·34	10·97	33·5	38·6	43·8
Ar	1·52	2·66	3·94	5·78	7·22	8·76	11·94	13·77	40·7	46·1
K	0·41	3·08	4·43	5·87	8·00	9·63	11·36	14·92		
Ca	0·58	1·15	4·91	6·45	8·09	10·49	12·32	14·25		
Sc	0·62	1·25	2·41	7·12	8·86	10·69	13·38	15·31		
Ti	0·65	1·30	2·64	4·14	9·62	11·55	13·57	16·56		
V	0·64	1·40	2·56	4·62	6·26	12·42	14·54	16·75		
Cr	0·64	1·61	3·08	4·91	6·93	8·66	15·40	17·71		
Mn	0·71	1·50	3·27	5·10	7·32	9·72	11·46	18·77		
Fe	0·75	1·59	2·89	5·39	7·61	10·11	12·80	14·54		
Co	0·75	1·67	3·27	5·10	7·90	10·49	13·28	16·37		
Ni	0·73	1·75	3·46	5·39	7·61	10·88	13·77	16·84		
Cu	0·74	1·92	3·66	5·68	7·99	10·49	14·25	17·52		
Zn	0·90	1·73	3·85	5·97	8·28	10·97	13·86	18·10		
Ga	0·58	1·97	2·98	6·16	8·66	11·36	14·34	17·62		
Ge	0·77	1·54	3·27	4·43	8·95	11·84	14·92	18·20		
As	1·01	1·92	2·69	11·55	14·83	18·29	23·9	27·5		
Se	0·93	2·02	3·27	4·14	7·03	7·89	15·46	19·45		
Br	1·13	1·83	3·46	4·81	5·78	8·38	10·01	20·2		
Kr	1·32	2·2	3·5							
Rb	0·39	2·63								

SUCCESSIVE IONIZATION ENERGIES OF SELECTED ELEMENTS

(Values expressed in MJ/mol)

Element	1	2	3	4	5	6	7	8	9	10
Sr	0·55	1·06	4·14							
Y	0·62	1·18	1·96							
Zr	0·66	1·35	2·31							
Nb	0·65	1·34	2·36							
Mo	0·70	1·38	2·60							
Ru	0·72	1·54	2·79							
Rh	0·74	1·73	2·98							
Pd	0·78	1·92	3·18							
Ag	0·73	2·05	3·47							
Cd	0·87	1·63	3·64							
In	0·55	1·82	2·72							
Sn	0·70	1·42	2·91							
Sb	0·84	1·67	2·41							
Te	0·87	1·84	2·93							
I	0·98	1·83	3·01							
Xe	1·17									
Cs	0·37	2·22	3·26							
Ba	0·50	0·90	3·56							
La	0·53	1·09	1·84							
Pt	0·86	1·80	2·79							
Au	0·89	1·92	2·89							
Hg	1·00	1·82	3·28							
Tl	0·58	1·97	2·88							
Pb	0·71	1·44	2·93							
Bi	0·84	1·63	2·51							

Bibliography

Addison, W. E. 1963. *Structural principles in inorganic compounds*. London: Longman.

Cartmell, E. and G. A. Fowles 1966. *Valency and molecular structure*, 3rd edn. London: Butterworth.

Cotton, F. A. and G. Wilkinson 1972. *Advanced inorganic chemistry*, 3rd edn. New York: Wiley-Interscience.

Dasent, W. E. 1970. *Inorganic Energetics*. London: Penguin Library of Physical Science.

Greenwood, N. N. 1964. Chemistry of the noble gases. *Education in Chemistry*, **1**, 176.

Heslop, R. B. 1964. Stoichiometric valency and oxidation number. *J.R.I.C.*, **88**, 30.

Ives, D. J. G. 1960. *Principles of the extraction of metals*. (Monographs for Teachers, No. 3). London: Royal Institute of Chemistry.

Moody, G. J. and J. D. R. Thomas 1964. Compounds of the noble gases. *J.R.I.C.*, **88**, 31.

Rudge, A. J. 1962. *Manufacture and use of fluorine and its compounds* (I.C.I. Ltd). London: Oxford University Press.

Samuel, D. M. 1972. *Industrial chemistry—inorganic*. (Monographs for Teachers, No. 10). London: Chemical Society.

Sharpe, A. G. 1968. *Principles of oxidation and reduction*. (Monographs for Teachers, No. 2). London: Royal Institute of Chemistry.

Shreve, R. N. 1967. *The chemical process industries*, 3rd edn. New York: McGraw-Hill.

Spice, J. E. 1975. *Chemical binding and structure*, 2nd edn. Oxford: Pergamon.

Wells, A. F. 1975. *Structural inorganic chemistry*, 4th edn. London: Oxford University Press.

INDEX

211

212

214